SOCIAL WORK AND COMMUNITY PRACTICE

SOCIAL WORK AND COMMUNITY PRACTICE

Sharon Duca Palmer, CSW, LMSW

*School Social Worker, ACLD Kramer Learning Center,
Bay Shore, New York; Certified Field Instructor,
Adelphi University School of Social Work,
Garden City, New York, U.S.A.*

Apple Academic Press

Social Work and Community Practice

© Copyright 2011*
Apple Academic Press Inc.

First Published in the Canada, 2011
Apple Academic Press Inc.
3333 Mistwell Crescent
Oakville, ON L6L 0A2
Tel. : (888) 241-2035
Fax: (866) 222-9549
E-mail: info@appleacademicpress.com
www.appleacademicpress.com

The full-color tables, figures, diagrams, and images in this book may be viewed at www.appleacademicpress.com

First issued in paperback 2021

ISBN 13: 978-1-77463-250-5 (pbk)
ISBN 13: 978-1-926692-86-9 (hbk)

Sharon Duca Palmer, CSW, LMSW

Cover Design: Psqua

Library and Archives Canada Cataloguing in Publication Data
CIP Data on file with the Library and Archives Canada

CONTENT

INTRODUCTION

Social work is a difficult field to operationally define, as it is practiced differently in many settings. It is a very diverse occupation and one that can be practiced in settings such as hospitals, clinics, welfare agencies, schools, and private practices.

The main goal of all social work practice is to assist the client to function at the best of their ability and assess what their needs are. Social workers help clients with problem-solving strategies, such as defining personal goals, focusing on what is necessary to make changes, and helping them through the process.

Social work is a demanding field and is often emotional draining. Many social workers have large caseloads, limited resources for their clients, and often work for relatively low salaries. But the personal rewards can be very satisfying.

The social work profession is committed to promoting social and economic policy though helping to improve people's lives. Research is conducted to improve social services, community development, program evaluation, and public administration. The importance of research in these areas is to examine variables that can be addressed in order to resolve issues. Research can lead to what is called "best practice". By utilizing "best practice", a social worker is engaging clients based on research that is intended to increase successful outcomes.

Social work is one of the most diverse careers available. Most social workers are employed by health care facilities and government agencies. These facilities can include hospitals, mental health clinics, nursing homes, rehabilitation centers, schools, child welfare agencies, and private practice.

Social work's interface with mental health promotion and the treatment of mental illness dates to the earliest roots of our profession. While many social workers provide mental health services in private practice settings, the majority of services are offered in community-based agencies, both public and private, and in hospitals and prisons. Social workers are the largest provider of mental health services, providing more services than all other mental health care providers combined. These workers also often provide services to those who are struggling with substance abuse.

Twenty-first century health issues are complex and multidimensional, requiring innovative responses across professions at all levels of society. Public health social workers work to promote health in hospitals, schools, government agencies and local community-based settings, making connections between prevention and intervention from the individual to the whole population.

In an ideal world, every family would be stable and supportive. Every child would be happy at home and at school. Every elderly person would have a carefree retirement. Yet in reality, many children and families face daunting challenges. For example, single parents struggle to raise kids while working. Teens may become parents before they are ready. Child social workers help kids get back on track so they can lead healthy, happy lives.

Rapid aging populations are expected worldwide. With the rapid growth of this population, social work education and training specializing in older adults and practitioners interested in working with older adults are increasingly in demand. Geriatric social workers typically provide counseling, direct services, care coordination, community planning, and advocacy in an array of organizations including in homes, neighborhoods, hospitals, senior congregate living and nursing facilities. They work with older people, their families and communities, as well as with aging-related policy, and aging research

In whatever subcategory they work, social workers help provide support services to individuals and communities by assessing their needs in order to improve the quality of life and overall well-being. This can lead to positive changes in people's environments, dignity, and self-worth. It can also lead to changes in social policy for those who are vulnerable and oppressed. Social workers change entire communities for the better.

There have been many changes emerging in the social work profession. The uses of the Internet and online counseling have been major trends. Some people are more likely to seek assistance and information first through the use of the Internet. There has also been a strong move for collaborating between professions when providing services in order to offer clients more options for success. Keeping up to date with best practice research, licensing requirements, continuing education, and professional ethics make this an exciting and challenging time to be a social worker!

— Sharon Duca Palmer, CSW, LMSW

Pathways into Homelessness: Recently Homeless Adults Problems and Service Use Before and After Becoming Homeless in Amsterdam

Igor R. van Laere, Matty A. de Wit and Niek S. Klazinga

ABSTRACT

Background

To improve homelessness prevention practice, we met with recently homeless adults, to explore their pathways into homelessness, problems and service use, before and after becoming homeless.

Methods

Recently homeless adults (last housing lost up to two years ago and legally staying in the Netherlands) were sampled in the streets, day centres and overnight shelters in Amsterdam. In April and May 2004, students conducted interviews and collected data on demographics, self reported pathways into homelessness, social and medical problems, and service use, before and after becoming homeless.

Results

Among 120 recently homeless adults, (male 88%, Dutch 50%, average age 38 years, mean duration of homelessness 23 weeks), the main reported pathways into homelessness were evictions 38%, relationship problems 35%, prison 6% and other reasons 22%. Compared to the relationship group, the eviction group was slightly older (average age 39.6 versus 35.5 years; p = 0.08), belonged more often to a migrant group (p = 0.025), and reported more living single (p < 0,001), more financial debts (p = 0.009), more alcohol problems (p = 0.048) and more contacts with debt control services (p = 0.009). The relationship group reported more domestic conflicts (p < 0.001) and tended to report more drug (cocaine) problems. Before homelessness, in the total group, contacts with any social service were 38% and with any medical service 27%. Despite these contacts they did not keep their house. During homelessness only contacts with social work and benefit agencies increased, contacts with medical services remained low.

Conclusion

The recently homeless fit the overall profile of the homeless population in Amsterdam: single (Dutch) men, around 40 years, with a mix of financial debts, addiction, mental and/or physical health problems. Contacts with services were fragmented and did not prevent homelessness. For homelessness prevention, systematic and outreach social medical care before and during homelessness should be provided.

Background

There is little evidence on good practice in caring for homeless people in the medical literature [1]. It has been reported that for homeless people life expectancy averages around 45 years, and that lack of access to health care services has too often proved a barrier to recovery, and, as a result, contributes to a downward spiral of deteriorating health and premature death [2]. Therefore, public services strategies should include homelessness prevention.

To prevent and reduce homelessness, strategies that address the general population and/or a targeted population could include housing benefits, welfare benefits, supplementary security income, supportive services for impaired or disabled individuals, programs to ameliorate domestic conflicts, programs to prevent evictions, discharge planning for people being released from institutions and (outreach) care programs for homeless populations [3,4]

Despite all these efforts and investments, and although there is broad consensus among policy makers and service providers that more resources and professional efforts should be dedicated to homelessness prevention, insufficient knowledge is available on how to accomplish this [3-5].

To identify starting points for homelessness prevention strategies in Amsterdam, the Netherlands, we have described in previous articles evictions from ones home as a major pathway of how people enter homelessness [6,7]. We demonstrated that evictions were a neglected public health problem. Despite knowledge about the underlying social and medical problems among households at risk, referrals to social and medical care are insufficiently used as a method to prevent eviction. Furthermore, we concluded that in Amsterdam nobody took the responsibility for the evicted households, predominantly due to rent arrears, whether they became homeless or not.

The absence of integrated social medical care results in a lack of assistance for recently evicted households, many of whom enter homelessness. Once homeless, people are responsible themselves in their search for specific services, organised alongside the mainstream service delivery system [8,9]. The lack of assistance for recently homeless people seems to be in concordance with a lack of knowledge of recently homeless people, related to their pathways into homelessness and their social and medical problems [3-5,9,10]

Objective of this Study

Regarding the lack of assistance for evicted households in Amsterdam, and contributing to the knowledge on recently homeless people and the development of prevention practice, for this study we tried to identify recently homeless adults, to explore 1) the pathways into homelessness, 2) the social and medical problems before and after becoming homeless, and 3) the contacts with social and medical services before and after becoming homeless.

Methods

Study Population

Included in our study were recently homeless adults defined as persons, 18 years and older, who lost their house for the first time during the last two years (between

April 2002 and April 2004) and who were legally staying in the Netherlands. The choice of the length of homelessness up to two years was intended to enhance the reliability of the information reported and to overcome problems of memory. To find locations to meet recently homeless adults, data on rough sleepers and visitors of day centres in Amsterdam were studied [11,12] Staff at one specific benefits provider for the homeless, at five day centres and at two emergency shelters were interviewed for information on their homeless visitors. After combining oral and written information, we decided to reach as many recently homeless adults as possible at locations recently homeless people tend to visit and where they could be approached for an interview. These were three gathering places for outreach soup distribution and popular street hangouts, one specific agency for benefits provision for the homeless, four emergency shelters and seven day centres with each over 450 visits a week. To keep a homogeneous sample, shelters for adolescents and families were not included. The study design did not need a process of ethical approval according to the Dutch Act on Medical Research.

In April and May 2004, interviews were conducted by ten undergraduate social science students. The students were familiar with approaching and interviewing homeless people. Interviewers underwent three training sessions on the process and quality of data registration, and all questionnaires were reviewed after the interviews. For every completed questionnaire students received twenty euros. Interviews lasted on average 45 minutes.

During a total of 40 occasions, at fourteen locations, between 4 and 38 homeless people were present at any moment (on average 25), of whom 125 homeless adults were eligible and participated in the study, by giving written consent for an interview and anonymous data analysis. Specific encouragement or incentives for homeless people to participate were not applied. None of the respondents were too intoxicated or too confused to be able to participate. During the interviews, on a separate list, the questionnaire number, a coded name and date of birth of participants were recorded to exclude doubling; two persons were interviewed twice and were excluded from analysis. Three questionnaires were excluded as the respondents were homeless for longer than two years. In total 120 questionnaires were included in the analysis.

Collected Data

Questionnaires for this study consisted of author-generated items. In consultation with city sociologists at the University of Amsterdam Department of Social Sciences, items of questionnaires used in follow up studies on rough sleepers were added [11]. Data were collected in a variety of areas addressing who, where, what, how and when questions following the process and antecedents of becoming homeless, self reported social and medical problems and contacts with social and medical services, before and after becoming homeless. Type of underlying problems chosen were based on the authors experience with providing outreach care to homeless people in Amsterdam over the last decade [13].

To find out pathways into homelessness, respondents were asked about their last housing condition and included composition of the household, type of housing, type of lessor, rent agreement and rent/income ratio. Demographics included sex, age and country of birth. For information on the social and medical problems before and after becoming homeless the following items were asked. Social problems were domestic conflicts (with household members, neighbours, landlords and/or services) and financial problems. For the latter data on financial debts, reasons for debts and type of creditors were collected. Medical problems included addiction to alcohol, drugs and gambling, mental health problems and physical health problems. Alcohol use could be scored as normal, excessive or extreme, according to the Garretsen scale [14]. Cocaine and/or heroine use could be more or less than 13 days a month; 13 days were chosen to exclude weekend users. Gambling could be absent or present. For mental problems no specific instruments or criteria were used for practical reasons. Respondents were asked if they felt depressed, fearful and/or confused. Physical problems and/or handicap were asked in an open question.

Service contacts, before and after becoming homeless, included social and medical services. Social services included social work, benefits agency, debt control agency, as well as shelters and day centres. Medical services included general practitioner, addiction service, mental health service and the GGD Municipal Public Health Service (safety net department and outpatient drug clinics) [13].

Study Assessments and Analysis

Statistical analyses were performed using SPSS 14.0 and were mainly descriptive. The pathways into homelessness are described. Demographics, problems and service use are described and compared between the three main identified pathways into homelessness. Differences in the characteristics and underlying problems among homeless people following the different pathways are compared using chi-square and Fisher-exact tests for categorical variables and Wilcoxon median test for continuous variables. To identify independent factors associated with the specific pathways, logistic regression analyses was performed using backwards selection based on the loglikelihood ratio. In addition, logistic regression analyses was performed to study factors independently associated with the main problems identified in each pathway.

Results

Housing Setting and Pathways into Homelessness

In table 1 the self reported housing setting and pathways into homelessness are shown. Before homelessness two thirds were living in a rented house. Thirteen respondents, out of the 120, mentioned never having lived independently; they had always been staying with family or friends. More than half had rented a house of a housing association (53%) and one third had rented privately (32%). The median rent price was 268 euros (range 0 – 1,000 euros), and the median gross salary was 809 euros (range 0 – 4,500 euros). Forty respondents had a rent/income ratio up to 30%, 33 up to 60%, 7 more than 60% and for 40 respondents this was not known.

Table 1. Self reported setting and pathways into homelessness

	n	%
Type of housing (n = 115)		
own house	75	65
stay with family, friends or other	27	23
prison	6	5
abroad	3	3
hospital	2	2
hostel	2	2
Type of tenant (n = 85)		
housing association	45	53
private rent	27	32
subletting	10	12
other	3	4
Pathways, how last housing lost? (n = 109)		
eviction	41	38
relationship problems (left or sent away)	38	35
after prison	6	6
other reasons	24	22
where last housing lost? (n = 117)		
Amsterdam	95	81
rest of the Netherlands	16	14
abroad	6	5
whereabouts after loss of housing? (n = 114)		
family or friends	42	37
rough sleeping	30	26
shelter for the homeless	25	22
squads and garage boxes	8	7
abroad	5	4
clinic	4	4
when homeless after loss of housing? (n = 98)		
immediately	56	57
1 day – 1 week	8	8
1 week – 1 month	8	8
1–3 months	13	13
3–6 months	8	8
> 6 months	5	5
how long currently homeless? (n = 120)		
< 1 month	11	9
1–3 months	11	9
3–6 months	42	35
6–12 months	30	25
12–18 months	16	13
18–24 months	10	8

120 recently homeless people in Amsterdam who lost their house for the first time between April 2002–April 2004.

When asked how respondents lost their last housing, answered by 109 re-spondents, the three main pathways were evictions (38%), leaving ones house or being send away by others due to relationship problems (35%) and other reasons (28%). Among 38 respondents who were homeless due to relationship problems, (of whom one third had a rent contract in their own name), 4 had left on their own initiative and 34 were sent away by household members (partner 22, parents 6 and roommates 6). Among other reasons, 6 mentioned they had lost their house while doing time in prison. Four out of five had become homeless in Amsterdam. After loss of last housing, 57% reported immediate homelessness, and 86% reported being on the streets within three months. The median length of homeless-ness was six months (23 weeks).

Demographics and Household Composition

In table 2 demographics and household composition related to pathways through which people became homeless are shown. Compared to the rela-tionship group, the eviction group was slightly older (average age 39.6 versus 35.5 years; p = 0,08), living single more often (p = 0,000), and belonged to one of the major migrant groups more often (p = 0.025). The total average age for both sexes was 38 years, the range for males was 18–67 years, and for females 19–50 years.

Table 2. Demographics and household composition related to pathways into homelessness

Demographics	total (n = 120)		eviction (n = 41)		relationship (n = 38)		other (n = 30)	
	n	%	n	%	n	%	n	%
sex								
male	105	88	38	93	33	87	26	87
female	15	12	3	7	5	13	4	4
Age in years								
18–29	39	33	8	20	13	35	10	33
30–39	30	25	15	38	11	30	8	27
40–49	26	22	8	20	7	19	6	20
50–59	14	12	6	15	4	11	3	10
60–67	9	8	3	8	2	5	3	10
Country of birth								
Netherlands	58	48	15	38	18	47	21	70
Surinam/Antilles/Morocco	24	20	13	33*	6	16	4	13
Other	37	31	12	30	14	37	5	17
composition of household								
single	46	44	26	63**	3	10	12	48
with parents	6	6	1	2	3	10	2	8
with partner	22	21	4	10	13	43	4	16
with partner and children	18	17	4	10	8	27	4	16
single parent	3	3	2	5	0	0	1	4
with other adult	10	9	4	10	3	10	2	8

120 recently homeless people in Amsterdam, who lost their house for the first time between April 2002–April 2004.
* p = 0.025; ** p = 0.000

Pathways and Problems

Self reported problems before homelessness related to pathways into homelessness are shown in table 3. Social problems were present in 81% of the total group, and medical problems in 76%. In the total group, before homelessness, almost two thirds (62%) had both social and medical problems (not in table 3). As expected, regarding pathways and social problems, the eviction group had significantly more often financial debts than the relationship group (p = 0,009). Logistic regression analyses showed that the only factor independently associated with financial debts were alcohol problems (OR 7.0 (95% CI: 2.0–25.0). Also in the relationship group almost half reported financial debts. The relationship group reported more domestic conflicts than the other groups (p < 0.001). Domestic conflicts were more common among those between 18–29 years and those 60 years and older, and among respondents not born in the Netherlands. Underlying social or medical problems were not significantly associated with domestic conflicts.

Table 3. Self reported problems before homelessness related to pathways into homelessness

Reported problems before homelessness	total (n = 120)		eviction (n = 41)		relationship (n = 38)		other (n = 29)	
	n	%	n	%	n	%	n	%
Social problems (n = 109)	88	81	35	85	32	84	21	70
financial debts	73	61	33*	81	18	47	18	62
domestic conflicts	55	46	18	44	29***	76	6	21
Medical Problems (n = 109)	83	76	32	78	30	79	21	70
Addiction total	57	48	24	59	17	45	10	35
alcohol ##	26	22	12***	29	5	13	4	13
drugs	37	31	12	29	15	40	6	21
cocaine	33	28	10	24	14	37	6	21
heroin	14	12	5	12	6	16	1	3
gambling	22	18	10	24	5	13	4	14
Mental problems	67	56	26	63	20	53	16	53
depressed	61	51	24	59	18	47	14	47
fearful	29	24	15	37	6	16	5	17
confused	23	19	8	20	6	16	7	23
Physical problems	26	22	7	17	10	26	8	28

120 recently homeless people in Amsterdam who lost their house for the first time between April 2002–April 2004.
according Garretsen scale [14]
*p = 0.009; **p < 0.001, ***p = 0.048

Regarding pathways and medical problems, the eviction group reported more extreme alcohol problems than the relationship group (p = 0.048). Drug problems, mainly cocaine use, tended to be more common in the relationship group compared to the eviction group, although not significantly. In all groups more than half reported mental health problems.

Not reported in table 3, among 73 respondents with debts, the main reasons for debts were loss of job and/or chronic shortage of income (49%), buying drugs (18%), gambling (10%), and other reasons such as fines, order by credit and health costs (23%). Of 22 gamblers, 16 respondents reported financial debts. Of

73 respondents with debts, 16 reported gambling. The majority of creditors were banks (35%), landlords (34%), energy companies (18%) and family members (9%). The median debt was 5,000 euros (range 400 – 400,000 euro). One person had left a mortgaged house leaving a 400,000 euros debt.

In table 4 problems before and after becoming homeless are shown. Financial problems before homelessness (61%) were not solved during homelessness, when even more respondents reported debts (68%). The overall addiction rate had decreased from 48% before to 20% after becoming homeless, due to less excessive and extreme use of alcohol, less use of heroin, and less use of cocaine. The self reported gambling rate decreased from 18% before to 3% after becoming homeless. During homelessness only a few individuals began substance use for the first time. Feelings of being depressed, fearful and confused were frequently reported before (56%) and after (63%) becoming homeless, and in both periods almost one quarter reported physical problems or a handicap.

Table 4. Self reported problems before and after becoming homeless

Reported problems	before n	%	after n	%
Social problems	88	81	82	68
financial debts	73	61	82	68
domestic conflicts	55	46	-	-
Medical Problems	83	76	89	74
Addiction total	57	48	22	20
alcohol*	26	22	9	8
excessive	15	13	3	3
extreme	11	9	6	5
drugs	37	31	17	14
cocaine	33	28	16	13
< 13 days per month	23	19	8	7
13+ days per month	10	8	8	7
heroin	14	12	6	5
< 13 days per month	6	5	2	2
13+ days per month	8	7	4	3
gambling	22	18	4	3
Mental problems	67	56	75	63
depressed	61	51	68	57
fearful	29	24	24	20
confused	23	19	22	18
Physical problems	26	22	29	24
Total social and medical	76	63	63	55

120 recently homeless people in Amsterdam who lost their house for the first time between April 2002–April 2004.
* according Garretsen scale [14]

Pathways and Service Use

The self reported service use for social and medical problems before and after becoming homeless is shown in table 5. Despite the fact that a combination of debts, addiction and/or mental health problems were often reported, contacts with social services were low and with medical services even lower. Among the contacts with medical services the general practitioner played a minor role.

Table 5. Self reported service use before and after becoming homeless

Service use	before		after	
	n	%	n	%
social services total	45	38	100	83
social work	26	22	44	37
benefit agency	18	15	49	41
debt control agency	17	14	12	10
shelters and daycentres	22	18	73	61
medical services total	32	27	32	27
general practitioner	6	5	13	11
addiction service	12	10	9	8
mental health service	13	11	7	6
GGD public health service*	13	11	14	12
all services	65	57	106	93

#120 recently homeless people in Amsterdam who lost their house for the first time between April 2002–April 2004.
*GGD safety net department and outpatient drug clinics [13].

Regarding pathways and social problems, the eviction group reported more contacts with debt control services than the relationship group (33/41 = 81% versus 18/38 = 47%; p = 0.009, not in table). Despite these contacts they did not keep their house. Regarding pathways and medical problems no significant differences in service use between pathway groups were found. Before homelessness, of 86 respondents who reported a medical problem, 47 did look for some sort of medical service and 39 did not feel the need. Reasons mentioned for not perceiving the need for medical support were e.g. "I don't need help," "I solve my own problems," "I don't have an addiction problem," "I don't see how they can help me," "I don't know where to go," "they ask too many questions" and "services are slow."

How Recently Homeless People Envision Better Services and their Biggest Dream

We asked recently homeless people about their ideas how to improve assistance. In general, the majority of respondents mentioned that they wished that the city provides a one stop comprehensive service for social and medical problems, active assistance for red tape and financial management, and fast tracking towards (guided) housing and jobs. Respondents said e.g.: "you need to be verbally strong to succeed at services," "social and financial support should be much faster," "I wish clear information where to go for what problem," "services should work together." Other answers were: "If I had help before I became homeless....," "I try to be nice, but they are rude," and "they should offer help for normal homeless people."

What is your biggest dream? Almost all wanted a house, a normal life with family contacts and/or a job. Respondents said e.g. "I hope they give me benefits in the future," "to see my daughter," "a safe place," "a house within a few months, and celebrate Christmas with friends at home." Other answers were: "that they do more for homeless people who do not take drugs," "I do not have dreams, I gave up hope a long time ago" and one man was dreaming of "a shower and clean clothes."

Discussion

For the homelessness prevention practice, we aimed to discover the sources of homelessness; defined as the factual pathway that leads to an (official) forced or voluntary displacement from ones home or facility. Therefore, we explored the pathways people took into homelessness and compared the characteristics, problems and service use per pathway taken. In our approach, we focus on the detection of underlying problems, that services should respond to, rather than exploring the reasons why the underlying problems exist. Knowledge of the characteristics and problems of people who follow different pathways into homelessness should contribute to timely detection of vulnerable people who might step into homelessness.

We identified 120 recently homeless people in Amsterdam to explore their pathways into homelessness, problems and service use, before and after becoming homeless. The main pathways into homelessness reported were evictions from ones home (38%), relationship problems that lead to leaving a home or being sent away by household members (35%), leaving prison (6%) and various other reasons (22%). These pathways into homelessness are consistent with those known in the literature [4-10,15,16]. However, the figures in this sample can not be compared with those found by others due to varying settings, definitions and methodology. For comparison, the factual pathways into homelessness, the key causes, underlying contextual factors and triggers need to be disentangled [4,5,9].

Not surprisingly, the characteristics of the recently homeless people in our study show more similarities than differences with those found among the majority of households at risk of eviction (due to rent arrears and nuisance), rough sleepers, shelter users and homeless adults visiting outreach medical care facilities in Amsterdam [6,11,17-19]. The profile of the majority of the homeless in Amsterdam is comparable with those in cities abroad [10,20-22].

In all pathway groups almost two thirds reported a combination of social and medical problems. Those who were homeless after eviction did belong to a major migrant group more often, were slightly older, were more often living single, had more financial problems and more alcohol problems, than the other groups. Those who were homeless due to relationship problems were slightly younger, had more domestic conflicts and tended to report more drug (cocaine use) problems, than the others groups.

Gambling, as a known source of debts and financial difficulties, was reported by 24% among those evicted and 13% among those who had lived with others. In Melbourne, Australia, before homelessness, among 93 older homeless men, gambling was reported by 46% among those who were living alone and 28% among those living with others [4,16]. In Amsterdam, gambling was hardly mentioned by employees of housing associations handling rent arrears and by employees in nuisance control care networks handling nuisance, when asked to report problems among households at risk of eviction [6]. Service providers should be alert for gambling problems among mostly single men at the brink of homelessness due to financial difficulties.

Furthermore, regarding medical support before homelessness, for all pathways, the general practitioner, as a gatekeeper for addiction, mental and physical health problems, played a marginal role in providing care, which was also found among households at risk of eviction in Amsterdam [7]. For those at risk of homelessness with silent and/or non-self perceived health needs, 39 out of 86 who reported a medical problem, a sharp decrease of home visits carried out by general practitioners might be unfavourable [8,23]. Specifically, if no alternative social medical care at home is provided, and lessons how to integrate care for those in highest need have to be learned in the streets [25]. Therefore, rent arrears and nuisance can serve as signals to explore underlying problems by outreach support [6,7].

After becoming homeless, most problems identified before homelessness were also reported to exist afterwards, except for substance use and gambling, which had decreased significantly. The fact that many recently homeless had sought social care and were willing and capable of placing their addiction more in the background, is an indication of the motivation within this group to turn their situation around. The addiction decrease could be an indication that in the first homeless period the scarce financial means are being used mainly for subsistence. This moment should be an entry point for service providers to actively guide the recently homeless towards rehabilitation. Although validated diagnostic mental health tools were not used, by often reporting mental health problems many respondents did not seem satisfied with their mental health condition and/or situation. For recently homeless people staying in the same shelters and day centres together with the long-term homeless might have a numbing effect on a positive attitude towards rehabilitation [26,27].

The strength of this study is that we had good access to key informants and the locations where recently homeless people tend to gather. We obtained a high response rate among the recently homeless people who were approached for an interview. This study involved two principal limitations. First, our data regarding medical problems were based on self-reported information. Specifically for psychiatric problems diagnostic or clinical instruments were not used, therefore data can not be compared with other studies. Furthermore, some respondents mentioned having trouble remembering the number of services they had used over time. Second, a random sample of the recently homeless could not be drawn since the duration of homelessness is not registered at day centres and shelters, and not for those not using these facilities. Following our experience with homeless care, we believe that the data are valid and can be generalised for the total recently homeless population in Amsterdam.

Homelessness Prevention Strategies

Scholars in Australia, England and the US have described multiple obstacles for homelessness prevention strategies and the evaluation of prevention programs [3-5]. Regarding causes of homelessness, most cases involved personal problems and incapacities, policy gaps and service delivery defects. Crane et al. found that vulnerable people were being excluded because health and welfare services did not have the responsibility or resources to search for people with unmet treatment or support needs [4,5]. Furthermore, evaluation of homelessness prevention programs are hampered e.g. by fragmented and provision driven data registration [3].

In Amsterdam, several strategies to prevent and reduce homelessness have been implemented, since our study was executed in 2004. The Amsterdam Welfare and Care department promotes an integrated approach by housing, social and medical services to take responsibility in actively assisting vulnerable citizens with unmet support needs. This strategy is in concordance with the wishes and dreams of the majority of the recently homeless in our study. Since 2007 service providers are being trained for this approach to learn how to explore problems and pathways towards shared assistance. Furthermore, with substantial national and local financial support, services are able to expand their activities. More guided living options in the social housing sector (75% of the total housing stock in Amsterdam) are being offered, more integrated one stop social medical service units will be build, and the number of beds in shelters, addiction and mental health care facilities are being increased [28].

Regarding the three pathways into homelessness of the recently homeless people in our study, we reflect and comment on the existing strategies in Amsterdam.

Eviction from One's Home was the main Source of Homelessness

Per year more than 1,400 households are being evicted in Amsterdam [6]. To decrease the number of evictions, the existing outreach networks respond to persistent rent arrears and nuisance, as signals to be picked up by housing associations and landlords, to be shared with social services. In response, during a house visit underlying problems, such as gambling and medical problems, and unmet support needs are being explored [6,7,28]. Based on our previous studies on evictions and current findings, we suggest that assistance should explicitly be applied to low income single men, with underlying financial problems, addiction and/or mental and/or physical health problems. As among these high risk men a mix of social and medical problems is to be expected, social and medical workers should be trained to systematically approach and guide the underlying problems to keep these men at home [6,7,25].

Relationship Problems that Lead to Leaving a House was the second Source of Homelessness

Prevention strategies might be difficult to design. However, underlying problems and service use are also prevalent among this high-risk group. Alertness of social and medical services could be the way to identify this high risk group for preventive actions. Services should know their clients and should (be trained to) be sensitive for signals of vulnerability. These signals should be detected with a few additional questions related to how a person is coping with daily living, household management, income and debts (alcohol, cocaine and gambling), and should actively be shared among disciplines [4,5,21]. In health care settings medical professionals, and the general practitioner in particular, do have the opportunity and responsibility to diagnose social disease (such as poverty and imminent homelessness), that intrinsically interacts with medical disease, and actively ask for social assistance in response [5,29].

Leaving Prison was the third Source of Homelessness, among Various other Reasons

In the Netherlands, when people stay in prison for a certain period of time welfare benefits are being terminated. Data on the number of people that did pay rent off welfare benefits before they went to prison are not being collected. Nor data on the number of people that lost their house during time in prison because nobody assisted in paying the rent at home, and, as a consequence, became homeless after leaving prison. However, in Amsterdam, vulnerable inmates and multiple offenders are actively being followed up and assisted to anticipate housing, income and care after prison [28].

Furthermore, to prevent long term homelessness, new arrivals in the homeless circuit, at places the homeless tend to gather, are actively being identified and fast tracked along social and medical services, as the motivation to turn their situation around is expected to be a crucial entry point towards rehabilitation. For this strategy, social and shelter services aim to converge their intake procedures in a central shelter unit, where (recently) homeless people can undergo a social medical assessment and be guided towards problem oriented housing and care. Among the services for the poor and underserved, the GGD Municipal Public Health Service is operating as the central field director to monitor strategies to further prevent and reduce homelessness in Amsterdam [28]. New evaluations should demonstrate whether the present situation has improved compared to our findings in 2004.

Conclusion

Among recently homeless adults in Amsterdam, the main pathways into homelessness reported were evictions, relationship problems and leaving prison. In all pathways, the recently homeless fit the profile of the majority of the total homeless population in Amsterdam: single men, around 40 years, with a mix of debts, domestic conflicts, addiction, mental and/or physical health problems. Regarding service use before becoming homeless, and regardless the pathway taken, more than half reported contacts with social and/or medical services that did not prevent homelessness. During homelessness only contacts with social work and benefit agencies increased, contacts with medical services remained low. For homelessness prevention, systematic and integrated social medical care before and during homelessness should be provided.

Competing Interests

The authors declare they have no competing interests. No external funding was provided for this research.

Authors' Contributions

All authors contributed to the conceptualisation of the paper. IvL contributed to the study design and implementation, and wrote the manuscript. MdW contributed to the study design and implementation, analysed the data and assisted in writing the manuscript. NK contributed to the manuscript design and assisted in writing the manuscript.

Acknowledgements

We thank L. Deben, MSc, PhD, and P. Rensen, MSc, former city sociologists at the University of Amsterdam, for information on rough sleepers and design of the questionnaire, and sociology students for interviews and data collection. We thank shelter and day centre staff for their hospitality, interviews and information on visitors. Professor A. Verhoeff, PhD, GHA van Brussel, MD and TS Sluijs, MPH, all with the GGD Municipal Public Health Service Amsterdam, for their contribution to the study during the preparation phase and comments on earlier drafts of the manuscript. We also thank SW Hwang, MD, MPH, University of Toronto, Division of General Internal Medicine, St. Michael's Hospital, Toronto, Canada, for advice and comments on the manuscript.

References

1. Hwang SW, Tolomiczenko G, Kouyoumdjina FG, Garner RE: Interventions to Improve Health of the Homeless: a systematic review. Am J Prev Med 2005, 29(4):311–19.

2. O'Connell JJ: Premature Mortality in Homeless Populations: A Review of the Literature. [http://www.nhchc.org/PrematureMortalityFinal.pdf]. Nashville, USA: National Health Care for the Homeless Council, Inc; 2005.

3. Shinn M, Baumohl J, Hopper K: The prevention of Homelessness Revisited. Analyses of Social Issues and Public Policy 2001, 95–127.

4. Crane M, Byrne K, Fu R, Lipmann B, Mirabelli F, Rota-Bartelink A, Ryan M, Shea R, Watt H, Warnes AM: The causes of homelessness in later life: findings from a 3-nation study. J Gerontol B Psychol Sci Soc Sci 2005, 60(3):S152–9.

5. Crane M, Warnes AM, Fu R: Developing homelessness prevention practice: combining research evidence and professional knowledge. Health Soc Care Community 2006, 14(2):156–66.

6. van Laere IRAL, de Wit MAS, Klazinga NS: Evictions as a neglected public health problem: characteristics and risk factors of households at risk in Amsterdam, in press.

7. van Laere IRAL, de Wit MAS, Klazinga NS: Evaluation of the signalling and referral system for households at risk of eviction in Amsterdam. Health Soc Care Community 2008, in press.

8. Plumb JD: Homelessness: care, prevention, and public policy. Ann Intern Med 1997, 126(12):973–5. Review

9. Anderson I, Baptista J, Wolf J, Edgar B, Benjaminsen L, Sapounakis A, Schoibl H: The changing role of service provision: barriers of access to health services for homeless people. [http://www.feantsa.org/files/transnational_reports/2006reports/06W3en.pdf]. Brussels: Feantsa, European Observatory on Homelessness; 2006, 8–10.

10. Schanzer B, Dominguez B, Shrout PE, Caton CL: Homelessness, health status, and health care use. Am J Public Health 2007, 97(3):464–9.

11. Deben L, Rensen P, Duivenman R: [The homeless at night in Amsterdam 2003]. Amsterdam: Aksant; 2003. [Dutch]

12. [Exploration of the provision of drop in day centres for the homeless in Amsterdam] Amsterdam Gemeentelijke Dienst Maatschappelijke Ontwikkeling (DMO) afdeling Maatschappelijke en Gezondheidszorg 2003. [Dutch]

13. van Laere IRAL: Outreach Medical Care for the Homeless in Amsterdam. Ambulatory Medical Team: the years 1997–2004. Amsterdam: GGD Municipal Health Service; 2005.

14. Garretsen HFL: [Problem drinking: prevalence, associated factors and prevention: theoretical considerations and research in Rotterdam]. In Thesis. Lisse: Swets & Zeitlinger; 1983. [Dutch]

15. Crane M, Warnes AM: Evictions and Prolonged Homelessness. Housing Studies 2000, 15(5):757–773.

16. Rota-Bartelink A, Lipmann B: Causes of homelessness among older people in Melbourne, Australia. Australian and New Zealand Journal of Public Health 2007, 31(3):252–8.

17. Buster MCA, van Laere IRAL: [Dynamics and problems among homeless people using shelters in Amsterdam]. Amsterdam: GG&GD; 2001. [Dutch]

18. van Laere IRAL, Buster MCA: [Health problems of homeless people attending the outreach primary care surgeries in Amsterdam]. Ned Tijdschr Geneeskd 2001, 145:1156–60. [Dutch]

19. Sleegers J: Similarities and differences in homelessness in Amsterdam and New York City. Psychiatr Serv 2000, 51(1):100–4.

20. Morrell-Bellai T, Goering PN, Boydell KM: Becoming and remaining homeless: a qualitative investigation. Issues Ment Health Nurs 2000, 21(6):581–604.

21. Goering P, Tolomiczenko G, Sheldon T, Boydell K, Wasylenki D: Characteristics of persons who are homeless for the first time. Psychiatr Serv 2002, 53(11):1472–4.

22. Fountain J, Howes S, Marsden J, Strang J: Who uses services for homeless people? An investigation amongst people sleeping rough in London. Journal of Community & Applied Social Psychology 2002, 12(1):71–75.

23. Berg MJ, Cardol M, Bongers FJ, de Bakker DH: Changing patterns of home visiting in general practice: an analysis of electronic medical records. BMC Fam Pract 2006, 7:58.

24. Allen T: Improving housing, improving health: the need for collaborative working. Br J Community Nurs 2006, 11(4):157–161.

25. van Laere IRAL, Withers J: Integrated care for homeless people – sharing knowledge and experience in practice, education and research: Results of the networking efforts to find Homeless Health Workers. Eur J Public Health 2008, 18(1):5–6.

26. O'Toole TP, Gibbon JL, Hanusa BH, Fine MJ: Preferences for sites of care among urban homeless and housed poor adults. J Gen Intern Med 1999, 14(10):599–605.

27. Daiski I: Perspectives of homeless people on their health and health needs priorities. J Adv Nurs 2007, 58(3):273–81.

28. [Off the streets: better care, less homelessness and less nuisance. Changes in service delivery for the years 2007–2010] Amsterdam: Gemeente Amsterdam, Dienst Zorg en Samenleven; 2007. [Dutch]

29. van Laere IRAL: Caring for homeless people: can doctors make a difference? Br J Gen Pract 2008, 58(550):367.

Promoting Chlamydia Screening with Posters and Leaflets in General Practice — A Qualitative Study

Elaine Freeman, Rebecca Howell-Jones, Isabel Oliver,
Sarah Randall, William Ford-Young, Philippa Beckwith
and Cliodna McNulty

ABSTRACT

Background

General practice staff are reluctant to discuss sexual health opportunistically in all consultations. Health promotion materials may help alleviate this barrier. Chlamydia screening promotion posters and leaflets, produced by the English National Chlamydia Screening Programme (NCSP), have been available to general practices, through local chlamydia screening offices, since its launch. In this study we explored the attitudes of general practice staff to these screening promotional materials, how they used them, and explored oth-er promotional strategies to encourage chlamydia screening.

Methods

Twenty-five general practices with a range of screening rates, were purposively selected from six NCSP areas in England. In focus groups doctors, nurses, ad-ministrative staff and receptionists were encouraged to discuss candidly their experiences about their use and opinions of posters, leaflets and advertising to promote chlamydia screening. Researchers observed whether posters and leaf-lets were on display in reception and/or waiting areas. Data were collected and analysed concurrently using a stepwise framework analytical approach.

Results

Although two-thirds of screening practices reported that they displayed post-ers and leaflets, they were not prominently displayed in most practices. Only a minority of practices reported actively using screening promotional materi-als on an ongoing basis. Most staff in all practices were not following up the advertising in posters and leaflets by routinely offering opportunistic screen-ing to their target population. Some staff in many practices thought posters and leaflets would cause offence or embarrassment to their patients. Distribu-tion of chlamydia leaflets by receptionists was thought to be inappropriate by some practices, as they thought patients would be offended when being offered a leaflet in a public area. Practice staff suggested the development of pocket-sized leaflets.

Conclusion

The NCSP should consider developing a range of more discrete but eye catching posters and small leaflets specifically to promote chlamydia screening in different scenarios within general practice; coordinators should audit their use. Practice staff need to discuss, with their screening co-ordinator, how different practice staff can promote chlamydia screening most effectively using the NCSP promo-tional materials, and change them regularly so that they do not loose their im-pact. Education to change all practice staff's attitudes towards sexual health is needed to reduce their worries about displaying the chlamydia materials, and how they may follow up the advertising up with a verbal offer of screening op-portunistically to 15-24 year olds whenever they visit the practice.

Background

Genital chlamydia is the most common sexually transmitted disease in Europe [1,2]. The English National Chlamydia Screening Programme (NCSP), first introduced in 2003, offers opportunistic screening to sexually active young people aged 15-24 to reduce prevalence of chlamydia to prevent ectopic pregnancy, pelvic inflammatory disease and infertility in men and women [3]. General practice is widely used by young people [4] and, therefore, provides an opportunity to raise awareness of, and provide, chlamydia screening.

Health promotion posters and leaflets produced by the Department of Health and NCSP have been available to general practice through local chlamydia screening offices since the launch of the NCSP. The NCSP leaflet was first produced in April 2003 and was based on the leaflet originally used in the chlamydia screening pilot based in several health care settings including general practice [5]. It was produced by the National Chlamydia Screening Programme Steering Group, which had GP representation, and was reviewed by other stakeholders. The leaflet was designed using a Department of Health (DH) format and to fit in with the ongoing DH Sexual Health Awareness campaign. It was designed to be used by all clinicians as part of the screening consent procedure. The posters were produced by the DH as part of a Sex Lottery campaign to be used in a range of Health Care settings. Additionally, many local areas produce their own promotional information materials. This reflects the devolved nature of the NSCP with much of the funding and responsibility for publicity and promotion at local level [3].

We and others have found that many general practice staff admitted that they are reluctant to discuss sexual health opportunistically in all consultations [6-10]. Health promotion materials may help alleviate this barrier but there are no published studies of how the NCSP promotional materials are being used in England. The objective of this study was to explore the attitudes of general practice staff to health promotional materials aimed at increasing uptake of chlamydia screening, how staff used them, and explore other promotional strategies to encourage chlamydia screening. This was part of a larger qualitative study exploring general practice staffs' knowledge of the chlamydia screening programme and strategies they have used or suggest to encourage increased chlamydia screening within the general practice setting.

Methods

Disaggregate data from the Health Protection Agency (HPA) Centre for Infections [11] were used to identify and rank general practices by their chlamydia screening rates of their 16-24 year old target population. So that we could obtain the opinions of a wide range of general practices and staff, 25 high and low screening general practices were selected, using criterion based (purposive) sampling [12], from six NCSP areas in England who were encouraging screening within primary care. A member of the research team approached the practices by telephone and letter. We conducted eight focus groups with high screening practices (defined as those screening greater than 10% of their 16-24 year old target population), ten medium screening practices (between 3-10%) and 15 low screening practices (less than 3%).

Two high and six low screening practices declined to participate due to time pressures or staff shortages. Participating practices were visited between November 2005 and April 2007 and included those in both urban and rural locations, with a mix of social class and ethnic populations. Although researchers did not know the practice screening rates before each visit, it was difficult to blind them to whether the practice was a high or low screener, as this usually became quite apparent during the discussions. Doctors, nurses, administrative staff and receptionists were invited to participate in the focus group. The focus group schedule used open questions which encouraged respondents to discuss candidly their experiences of the chlamydia screening programme. As part of the focus group, we sought information about their use and opinions of posters, leaflets and advertising to promote chlamydia screening. Researchers observed whether posters and leaflets were on display in reception and/or waiting areas when they visited the practice and recorded field notes following each focus group. Data were collected and analysed using a stepwise framework analytical approach [13]. Focus groups were audio-taped then transcribed and checked for accuracy against the tapes. EF and CMcN used QSR NVivo software (QSR International PTY Ltd. Melbourne http://www.qsrinternational.com) to identify codes, categories and themes from the data, using an inductive approach. This approach was used as we wanted to be open to using the depth and breadth of data collected to show the opinions and behaviour of the whole general practice team. Themes were then discussed at a project meeting and agreed by all the authors. If there were any disagreements, the text was re-examined and a consensus reached. Data collection and analysis occurred concurrently and we continued, through purposive sampling, to enrol and visit practices to enrich the data [12]. The transcripts were then re-analysed by EF, using word-search, to ensure that any themes within the framework were not missed. The relationship between the practice screening rates (high, medium and low) and the data was examined. Quotations chosen demonstrate the different categories of data. These quotes were chosen as they highlighted the diversity of opinion in both high and low screening practices and differences between them.

Ethical approval was obtained from the Multi-Research Ethics Committee for Scotland (No. 4/MRE10/41). Local research governance approval was obtained from the relevant Trusts. Information sheets were sent to the practices at least two weeks before the focus group and all staff gave informed written consent and were assured anonymity.

Results

156 health care staff from 25 practices (participants per focus group 2-20; median 6) from urban and rural areas across England participated in the focus groups. These comprised 72 GPs, 46 nurses; eight practice managers; 23 receptionists/administrators and seven others.

Use of Chlamydia Screening Posters in General Practices

Major Themes

Use of Posters

Two thirds (16) of all screening practices said they had posters advertising chlamydia in their practices, which were either on the doors of their consulting rooms, in their waiting rooms or in patients' lavatories.

> *We also had [chlamydia] posters up around and also posters on our notice board outside and... it's on our LED. [electronic sign in reception] (Nurse FG14 medium screening practice)*

> *There's [general] posters throughout the surgery isn't there and in the passageways especially down by the nurses end. I don't think the age group is on them. There was a [chlamydia] poster up, but it has been taken down. (Receptionist FG8 low screening practice)*

However, researchers observed that most posters displayed in general practices were aimed at elderly people or promoted immunisation; very few had chlamydia posters in communal areas.

> *The difficulty with it, we tend to use the posters in the short term campaigns. We could do a campaign for an evening a month but then because of the wall space and everything we have to do. I mean we used to have lots of all sorts of [different] posters, it was too messy and too much information to read. (GP and nurse FG18 low screening practice)*

Many staff thought posters caused offence. Several practices had to deal with complaints from older people about posters advertising chlamydia screening in the waiting room or lavatories and patients had either removed the poster themselves or had asked for it to be taken down.

> *We put up a poster on how to do it [take a specimen] in the toilets that got taken down.*

We didn't have very much advertising [about chlamydia] because that actually upset quite a few patients. Especially the elderly, they were quite upset with our advertising. So we had to take it down and they said they didn't really want it in their face when they were sitting in the waiting room. We had quite a few people complain. We also had a piece on safe sex as well. And I think two people found that quite offensive, so we had to redo the advertising side. (GP FG2 medium screening practice respondent 1 & 3)

One low screening practice said there had been posters in nurses' rooms but these posters had been taken down and not replaced when the practice was refurbished. Many practices were concerned that posters may lose their impact if left on display.

My only problem is leaflets and posters go up and they become part of the scenery nobody takes any [notice] not a lot of impact. I think what we have to do would be [for] two weeks in a year, to have an impact on [chlamydia], have loads of posters up just for that [time]. (Practice nurse FG4 low screening practice)

A few low screening practices reported other priorities for wall space and did not wish to prioritise one disease over another.

We've got to be a bit sensitive. We've got the whole practice populations' needs [to think about] and I think we're not just here to deal with chlamydia we're here to deal with everything and things should be targeted equally. We've got a notice in each of our rooms on chlamydia so we have given it more space than some other diseases such as diabetes or asthma. We do try to give reasonably equal space don't we? (Practice nurse FG17 low screening practice)

A few professionals in low screening practices admitted they had not seen NCSP posters and were unaware of where they could access chlamydia screening health promotional materials.

I mean is it accessible [to us], or do we have to get our own literature? (Practice nurse FG11 high screening practice)

Use of Chlamydia Leaflets in General Practices

Major Themes

Most of the low screening practices had no chlamydia leaflets evidently on display when the researchers visited. Nineteen screening practices said they either had leaflets available in reception, waiting or consulting rooms, or included in self-sampling packs.

We leave leaflets in reception, and the primary care team produce a leaflet with all the different services that are available. Cards and a leaflet. We use it opportunistically for all our 16 year olds; that's part of the normal consultation. (Nurse FG3 high screening practice)

Displaying the leaflets in the waiting area for patients to help themselves was practices' main strategy for use.

They [the leaflets] are quite good. There's leaflets that we keep in the waiting room and by the reception. (Practice nurse FG21, medium screening practice)

[We have] leaflets in waiting rooms (GP)

...and we do have leaflets in our rooms, sometimes we have the little stand with 'what is chlamydia' and what you can do. (Practice nurse FG26 low screening practice)

Minor Theme

No recognition of leaflets: A few health professionals admitted they couldn't remember seeing the NCSP leaflets.

The big screen one I don't know what it is. Is that the coloured one?

Could be?

....pass. (Nurse and two GPs FG 23 medium screening practice)

Facilitator: So what do you think about the chlamydia leaflets?

I can't even remember what they look like.

Sorry I don't know what they look like.

they're quite a colourful thing…to be honest to you no recollection that's it

(GP, nurse and receptionist FG24 low screening practice)

Staff Opinions about Chlamydia Screening Leaflets

Major Themes

Practices were Enthusiastic About the Leaflets

Most practices were generally enthusiastic about the NCSP leaflets. Fourteen practices thought that the NCSP leaflet was user friendly, featured different ethnic groups, was easy to read and very informative.

The leaflets for the patient are absolutely superb, very self-explanatory; they are quite small and very necessary because they're all waiting to go in to see the doctor. Its nice print, it's nicely put and nice to read it's so easy to explain. (Nurse FG11 high screening practice)

In this area it is important to have more than white faces on a leaflet, that's gone down well.

(Nurse FG15 low screening practice)

Several practices reported that they used the leaflets as part of their consent procedure for screening.

I think it's excellent... it's very informative, but it's easy to read as well, which is quite important. We're supposed to [give it to patients] because its part of their consent procedure; consent is based on [the] leaflet. (GP FG3 high screening practice)

Leaflets May Cause Offence

However, staff in several low and a few medium screening practices thought that giving young patients a leaflet would cause patients embarrassment, resentment or offence.

A patient might think are they picking on me? Why would they think I might have chlamydia? So I don't think it would go down so well here at the moment... sometimes they may be resentful as well. (Nurse FG15 low screening practice)

Of these a few low screening practices thought that the leaflets' style was condescending to young people and provided too much general information and omitted to advise on how often they should be screened.

They're fine, I think the actual main leaflet is fine, but I think the one with boys and girls on...I think it's a bit condescending really, I think they're actually quite simplistic, so its easy to follow, but boys and girls, we're talking about people who are sexually active! I think it's a bit insulting putting boys on one and girls on the other personally. (Nurse FG17 low screening practice)

Other professionals in a few low screening practices thought that a chlamydia leaflet (especially the brightly coloured one) was a label for young people who may not wish to be seen with it or who would be offended by the suggestion that they may have chlamydia.

I think one of the reasons why I don't always hand out leaflets, is that very often they [the target group] come in without a bag or anything and if they are walking around with this [chlamydia leaflet] its just the fact that it says 'chlamydia' everyone look at what I've [got], we're trying to normalise it but they often come in with just a jacket or a tiny purse.

That's a good point, quite a lot of them [leaflets] end up left on the desk and then we find them when they've gone; I wouldn't walk out of the GP surgery with that in my hand either... What is it saying and yet you've only come in to have your travel vacs? (GP and nurse FG01 high screening practice)

Several thought leaflets should be concealed in envelopes. And a few other medium and low screening practices, following complaints from older people about chlamydia leaflets being highly visible in reception areas, had already put them in brown envelopes.

If the leaflet was in a brown envelope and you target an age group you could write along it hope you don't mind, but would you like to read this? I think they are a bit intrusive, because it would seem to some people that you were targeting them because you thought they might have a problem and I don't think that's right. (Receptionist FG8 low screening practice)

I think that's why we put them in brown envelopes because we wanted to reduce the older patients' concerns about it so we had to think about the other patients as well. Several, elderly ladies said it's too much in my face when I walk in, and I don't want to see this sort of thing. (FG2 medium screening practice)

Leaflets were too Bulky

A few of all the practices thought chlamydia leaflets were bulky and did not fit easily into a jeans pocket, so leaflets were often left on the reception desk. These professionals suggested that the information could be given in a more concise form (bullet points) or be credit card sized and that the cover could be more discreet.

I mean this is the only one I've got now... [showing small card], because it's not too bulky [to go] in a pocket. I agree totally absolutely fantastic, especially for younger people, because it doesn't contain [too much], you know [you've] got to sit there and read it. It's bullet points straight to the point facts given, no hassle with that. (Nurse FG9 high screening practice).

Minor Themes

Leaflets should be Translated

Some high and medium screening practices with high ethnic populations thought that the leaflet should be translated into different languages as patients may not understand written English. However some concern was expressed about the cost implications for the PCT, particularly in some locations where there were many different ethnic populations. One medium screening practice with a high ethnic population commented that pictorial leaflets should reflect the multi-cultural society that could be affected by chlamydia.

Because people come with different languages, if you say some [thing] even if it's in simple English they would not understand the meaning of it. Preferably I would like to have them in as many languages as possible, but it's not really viable from the PCT point [of view]. They cannot have an enormous amount of languages just for chlamydia screening, there's other things as well. (GP FG25 medium screening practice)

It's all written in English. I've obviously got doubts about the value of translations with all different languages but I think they are required for our population.

...very expensive I think the cost to do it [the leaflet translation]. (2 GPs FG5 high screening practice)

Other Strategies for the Use of Chlamydia Leaflets by General Practices

One high screening practice reported that clinicians had given patients a chlamydia leaflet whenever they attended for any consultation and this had resulted in a good screening.

Making sure the leaflets are available and being a bit more pro-active about things... because we found that really worked before, we had a good uptake. (Nurse and GP FG11 high screening practice)

Furthermore, receptionists in a few high screening practices proactively gave patients in the at-risk age group a chlamydia leaflet to read when they booked in to see a doctor or nurse and encouraged patients to ask for a chlamydia screen.

The leaflets, we hand it over and say would you like to read this while you're sitting waiting to see your doctor, if you would like to take part, speak to the doctor when you go in. (Receptionist FG5 high screening practice)

Sending Leaflets by Post

A few practices thought that patients could be sent a leaflet, with an invitation to attend for chlamydia screening, by post and this would prompt patients to consult the practice for a screening test. These professionals thought that if patients were aware that they were all being targeted in this way it would encourage more young people to attend.

Well we send out a letter to all the teenagers inviting them to the clinic and telling them what time, and it does say that we do tests for STIs. We're not specifically sending them information about chlamydia actually.

Maybe we should be enclosing a [chlamydia] information leaflet...most people coming in we are mentioning it anyway. (2 GPs FG6 medium screening practice)

I think you've got to do it by post. You send a leaflet through the post saying that there is a [chlamydia screening] programme available; these are the people who are at risk; you may wish to consult; these are the complications if it's not treated; you may wish to consider it. (GP, FG26 low screening practice)

A few medium screening practices thought that mobile phone texts may be a better way of encouraging young people to come forward for screening rather than sending letters to patients' homes.

I don't know if this national advertising works, [use] the text via mobile, that would be a good point, they've all got mobile phones. (GP FG13 medium screening practice)

Other Strategies for Increasing Awareness of Chlamydia Screening in General Practice

Major Themes

National Television Advertising

Many professionals in most of all the practices thought that NCSP needed a national advertising campaign either on prime time television or radio or included as an item in a leading soap opera. A few high and medium screening practices thought this high profile approach might be a better way to raise awareness of chlamydia screening with a multi-ethnic population.

I mean what about a national campaign on prime time television, what I'm saying [is a] campaign on television that you can go to your GPs and be tested or [get a] self-test kit. Definitely I think better advertising would be a better use of time and energy really. (Nurse FG13 medium screening practice)

What about a national campaign [on] prime time television. (Nurse FG4 low screening)

I think rather than this leaflet thing you should really increase your publicity in the media probably because the average person, no matter what his language will see television, and the advertisement should be at a time when they see that Eastenders, Big Brother and whatever you think, because a lot of people will understand that language. If a youngster in Eastenders has a problem then everybody will come. (GP FG25 medium screening practice)

Several practices suggested that including information on chlamydia screening and screening sites on practice websites would help to raise the profile of screening.

I think maybe advertising on the website and places that young people were going to look and read, as well as having the whole list of where they can get access to everything, not just one service. (Nurse FG19 low screening practice)

Several respondents commented that the promotional materials were also an important reminder for clinicians.

Keep reminding us, keep bringing it up in meetings, trying to change the way you advertise and put the message across, always trying to think of new ideas.

Oh, no no we were saying chuck around a few of the posters.

Also reminding clinicians, I think you have to keep on top of it really because I know our numbers dipped for a couple of months. (FG 2 medium tester respondent 3 and 4)

Discussion

Key Findings

Although two-thirds of practices reported they were displaying posters and leaflets they were not prominently displayed in most practices. Only a minority of practices reported actively using the posters and leaflets on an ongoing basis, to raise the visual profile of chlamydia screening, in reception, waiting areas, lavatories and consulting rooms. Although about half the practices were using posters and leaflets, many of these had low rates of screening and were not backing up the advertising by routinely offering opportunistic screening to their target population. However, the study design does not allow us to say that the use of promotional materials alone will increase screening rates. It was interesting that some staff in many practices thought using posters and leaflets would cause offence or embarrassment to their patients. Staff thought this could be decreased by using more discrete posters and smaller leaflets or by using envelopes. Distribution of chlamydia leaflets by receptionists was thought to be inappropriate by some practices, as they thought patients would be offended when being offered a leaflet in public waiting areas or reception. Practices suggested other ways to promote chlamydia screening including a national advertising campaign and sending letters to patients' homes.

Other Work in this Area

There is a paucity of literature exploring professionals' views about promotional posters and leaflets for chlamydia screening. Several studies exploring the use of promotional materials in other sexual health areas found that young patients are less likely to read posters than older patients [14-16]. These authors found that the public nature of waiting rooms and reception areas may inhibit patients collecting materials covering sexual health. Leaflets have been given to patients by GUM clinic receptionists to promote HIV testing [14,15]. Ivens and Sabin found that although patients' knowledge of HIV increased they were not more likely to accept a test compared to those offered a verbal discussion [15]. A recent general practice postal survey also shows that use of promotional materials covering sexually transmitted infections (STIs) is low in Australia—only one-fifth of GP respondents reported that they displayed posters covering STIs in their waiting room [17].

Andersen found a poor response to a multiple media campaign in Denmark which aimed to encourage chlamydia screening in young people [18]. Posters and leaflets in health, education and recreation centres, an internet web page, radio, television and newspaper interviews, were used to encourage young people to request a test kit for chlamydia. It resulted in a large proportion of requests for test kits from those ineligible to receive them [18], which indicates that leaflets need to be targeted at the at-risk population. This was done by several of the practices in our study.

It has previously been found that posters alone do not change the patient-professional interaction. In a US study, although 60% of patients noticed a poster campaign inviting patients in a US family practice clinic to discuss weight loss, the posters did not increase the proportion of patients reporting a change in patient-physician conversations about weight loss [19]. This work suggests that posters alone will not increase uptake of any health intervention without the willingness of health professionals to follow-up the advertising with offers to participate in screening.

Edwards et al's 2003 systematic review of communicating individual risk in screening programmes suggests that communication understanding is associated with higher uptake of tests, although none of the studies included in this review used posters or patient leaflet [20]. The interventions described were risk appraisal questionnaires, or tailored printed materials and counselling. These authors point out that further evaluation of strategies is needed to promote informed decision making and increase uptake of screening tests. O'Connor et al's 1997 systematic review found that decision aids were better than usual care for patient facing screening or treatment decisions [21]. Written decision aids supported patients' decisions by making them feel better informed [21]. This study points out that decision aids increased preferences for some interventions (for example Hepatitis B vaccinations) but not others (dental surgery) through increasing patients' knowledge but variations exist in behaviour to accept screening or treatment.

Strengths and Weaknesses

This is the first study to elicit the views of general practice staff, using qualitative methods, about NCSP chlamydia posters and leaflets. We expect that other general practices would show similar attitudes and beliefs to those demonstrated by the staff in this study. We used a focus group approach. Although interviews may have allowed more junior members of the team and non clinical staff to speak more frankly about their views, we decided on the focus group setting as this is the usual setting for practice meetings and how the chlamydia screening coordinators approach the practice to discuss screening. Our perception was that all staff were given the opportunity to vocalise their attitudes to screening. The questions covering promotional materials formed part of the longer focus groups and, therefore, we may have obtained even more detailed data if we had concentrated on a single area, but it is likely that practices and many staff may have declined to participate if covering just this narrower topic. Although specific open questions were used about posters and leaflets, we did not show participants the NCSP leaflets or posters. EF has nursing and research training and has no direct involvement in the NCSP, so this allowed her to approach the work without any preconceptions. We were not able to obtain patients' views of NCSP promotional materials, as it was outside the scope of this study, but is the focus of future work. The study design did not allow us to say whether the use of promotional materials will increase chlamydia screening. As part of a multifaceted intervention, promotional materials were successfully used to increase chlamydia screening in North Carolina, USA [22].

Implications of this Research

Implications for the NCSP and Chlamydia Coordinators

The NCSP should consider developing specific posters and leaflets promoting chlamydia screening that are suitable for the general practice waiting room. The posters need to be eye-catching for young people but not offensive to the elderly or those with young children. Leaflets need to be developed in different formats to suit different scenarios in the general practice setting. Credit card sized discrete leaflets would be more suited to receptions and waiting room areas, whereas longer leaflets are more suitable for clinicians to distribute as screening is offered. As staff suggested, in areas of high ethnicity the NCSP should consider leaflets in different languages and leaflets should inform young people how often they should be screened. Chlamydia coordinators need to promote and audit the use and display of chlamydia screening promotional materials more actively across all general practices registered with the chlamydia screening programme. Coordinators need to emphasise to practice staff that leaflets and posters need to be followed up with a verbal offer of screening opportunistically whenever a 15-24 year old visits the practice. All practice staff will need training on how to approach this.

In summary, the NCSP and coordinators of the service at PCT level should:

- Make posters more suitable for a general practice setting
- Make posters acceptable to elderly patients who may also view them
- Make leaflets less obviously about chlamydia so that they can be given to or picked up by patients without embarrassment
- Produce pocket or credit card sized leaflets
- In areas of high ethnicity, consider leaflets in different languages, or use more pictorial messages
- Audit use and display of posters and leaflets
- Ensure leaflets cover how often young people should be screened
- Educate practice staff on how to follow up leaflets and poster use with offer of a chlamydia screen

Implications for Practices

The NCSP needs to raise the profile of chlamydia screening in general practice. Practice staff need to discuss, with their screening coordinator, how the practice can promote chlamydia screening most effectively using promotional materials. The posters need to be displayed more prominently, so that they are more visible to young people, and practices should consider how they can increase distribution of leaflets through receptionists and clinicians. Practices need to review their use and display of sexual health promotion materials at least six monthly, as many professionals recognised that these lose their impact if not changed regularly.

Education to change practice staffs' attitudes towards sexual health is needed to reduce their worries and possible prejudices about displaying the chlamydia materials and following up with screening offers. A behavioural intervention approach using the Theory of Planned Behaviour [23] might be appropriate to address these issues. The promotional materials will be part of this approach to help normalise screening within the practice setting and make it more acceptable to staff and patients to offer screening opportunistically. Staff attitude to screening can also be influenced by education about the epidemiology and sequelae of chlamydia and the value of screening. Role play, videos or IT based materials can be used to increase practice staffs' confidence to be able to follow up poster displays and leaflets with verbal offers of chlamydia screening whenever they see a 15-24 year old patient. The education will need to be tailored for clinicians or the different staff; that is health care assistants and receptionists.

In summary, general practice staff should:

- Increase awareness of all staff of NCSP and the posters and leaflets available
- Display posters and leaflets more prominently
- Display leaflets where they can be easily picked up by young people
- Review display of leaflets and posters regularly, e.g. six monthly
- Agree on when leaflets should be offered to patients by receptionists
- Agree on when leaflets should be offered to patients by health-care assistants
- Agree on when leaflets should be offered by patients by nurses and doctors
- Order more discrete posters
- Order smaller more discrete leaflets
- Consider chlamydia envelopes if leaflets are not considered discrete enough for patients
- Undertake role play or video based education on how staff can offer leaflets
- Undertake role play or video based education on how staff can follow-up poster advertising in waiting room and patient leaflets with a screening offer
- Consider postal invitations for chlamydia screening

Conclusion

In conclusion the use of chlamydia screening posters and leaflets are not being optimised within most general practices. The NCSP posters need to be made eye-catching to the target group but acceptable to other patients visiting the practice. Leaflets need to be more discrete so that young people are happy to pick them up and read them. Clinicians need to follow-up poster advertising and leaflets with an offer of a chlamydia screen; video or role play based education may help them feel more confident to do this. To maintain their impact posters and leaflets need to be moved or changed six monthly and their use should be audited.

Competing Interests

Dr Cliodna McNulty writes the HPA Diagnosis of Chlamydia Quick Reference Guide for General Practices.

Isabel Oliver and William Ford-Young are members of the English National Chlamydia Screening Advisory Group; Sarah Randall is a former member.

Authors' Contributions

EF, RHJ, IO, SR, WFY, PB and CM contributed to the study design and writing of the protocol. EF, RHG & CM undertook the focus groups and analysed the data. EF and CM drafted the manuscript. RHJ, IO, SR, WFY and PB had input into the manuscript. All authors read and approved the final manuscript.

Acknowledgements

The study was funded by the Medical Research Council Grant No. G0500126.

Thank you to all the practice staff for participating in the study. Thanks to the Medical Research Council for funding the study and to the NCSP and Lynsey Emmett for providing us with screening data. Sue Starck, Allison Bates and Ji-yoon Knight are given grateful thanks for transcribing the focus groups. Thank you to Jill Whiting for her help with the grant and ethical applications, organising the focus groups and steering group meetings, and editing the paper.

References

1. McClure JB, Scholes D, Grothaus L, Fishman P, Reid R, Lindenbaum MD, Thompson RS: Chlamydia screening in at-risk adolescent females; an evaluation of screening practices and modifiable screening correlates. J Adolesc Health 2006, 38:726–733.

2. Scholes D, Stergachis A, Heldrich FE, Andrilla HJ, Holmes KK, Stamm WE: Prevention of pelvic inflammatory disease by screening for cervical infection. N Engl J Med 1996, 334:1362–66.

3. National Chlamydia Screening Programme Steering Group: Maintaining Momentum. Annual Report of NCSP in England 2006/07. Health Protection Agency, London; 2007.

4. Salisbury C, Macleod J, Egger M, McCarthy A, Patel R, Holloway A, Ibrahim F, Sterne JAC, Horner P, Low N: Opportunistic and systematic screening for chlamydia: a study of consultations by young adults in general practice. Brit J Gen Pract 2006, 56:99–103.

5. Department of Health: Chlamydia - you may not know you have it. Department of Health 2003. DOH 40149

6. McNulty CAM, Freeman E, Oliver I, Ford-Young W, Randall S: Strategies used to increase chlamydia screening in general practice: a qualitative study. Pub Health 2008, 122:845–56.

7. Cook RL, Wiesenfeld HC, Ashton MR, Krohn MA, Zamborsky T, Scholle SH: Barriers to screening sexually active adolescent women for: a survey of primary care physicians. J Adolesc Health 2001, 28:204–10.

8. Boekeloo BO, Snyder MH, Bobbin B, Burstein GR, Conley D, Quinn TC, Zenilman JM: Provider willingness to screen all sexually active adolescents for chlamydia. Sex Transm Infect 2002, 78:369–373.

9. Novak DP, Karlsson RB: Simplifying chlamydia screening: an innovative Chlamydia trachomatis screening approach using the internet and a home sampling strategy; population based study. Sex Transm Infect 2006, 8:142–47.

10. van Bergan J, Götz HM, Richardus JH, Hoebe CJPA, Broer J, Coenen AJT, for the PILOT CT study group: Prevalence of urogenital Chlamydia trachomatis increases significantly with level of urbanisation and suggests targeted screening approaches: results from the first national population based study in the Netherlands. Sex Transm Infect 2005, 81:17–23.

11. National Chlamydia Screening Programme Steering Group: New Frontiers Annual Report of NCSP in England 2005/06. Health Protection Agency, London; 2006.

12. Ritchie J, Lewis J, Eds: Qualitative research practice. London. Sage; 2004.

13. Ritchie J, Spencer L: Qualitative data analysis for applied policy research. In Analysing Qualitative Data. Edited by: Bryman A, Burgess RG. London. Ratledge; 1994:173–94.

14. Das S, Huengsberg M, Radcliffe K: Impact of information leaflets on HIV test uptake amongst GUM clinic attendees: an update. Int J STD & AIDS 2004, 15:422–423.

15. Ivens D, Sabin C: Providing written information on HIV testing improves patient knowledge but does not affect test uptake. Int J STD & AIDS 2006, 17:185–188.

16. Ward D, Hawthorne KB: Do patients read health promotion posters in the waiting room? A study in one general practice. Brit J Gen Pract 1994, 44:583–585.

17. Khan A, Plummer D, Hussain R, Minichiello V: Preventing sexually transmissible infections in Australian general practice. Int J STD & AIDS 2008, 19:459–63.

18. Andersen B, Ostergaard L, Moller JK, Olesen F: Effectiveness of mass media campaign to recruit young adults for testing of Chlamydia trachomatis by use of home obtained and mailed samples. Sex Transm Infect 2001, 77:416–418.

19. Stephens GS, Blanken SE, Greiner KA, Chumley HS: Visual prompt poster for promoting Patient-Physician conversations on weight loss. Annals of Family Medicine 2008, 6(Suppl):33–36.

20. Edwards A, Unigwe S, Elwyn G, Hood K: Effects of communicating individual risks in screening programmes: Cochrane systematic review. BMJ 2003, 327:703–707.

21. O'Connor AM, Rostom A, Fiset V, Tetroe J, Entwhistle V, Llewellyn T, Holmes Rovner M, Barry M, Jones J: Decision aids for patients facing health treatment or screening decisions: systematic review. BMJ 1999, 319:731–34.

22. Shafer MA, Tebb KP, Pantell RH, Wibbelsman CJ, Neyhaus JM, Tipton AC, Kunin SB, Ko TH, Schweppe DM, Bergman DA: Effect of a clinical practice improvement intervention on chlamydia screening among adolescent girls. JAMA 2002, 288:2846–52.

23. Ajzen I: The theory of planned behaviour. Organ Behav Hum Decis Process 1991, 50:271–87.

Evaluation of Effectiveness of Class-Based Nutrition Intervention on Changes in Soft Drink and Milk Consumption Among Young Adults

Eun-Jeong Ha, Natalie Caine-Bish, Christopher Holloman and Karen Lowry-Gordon

ABSTRACT

Background

During last few decades, soft drink consumption has steadily increased while milk intake has decreased. Excess consumption of soft drinks and low milk intake may pose risks of several diseases such as dental caries, obesity, and

osteoporosis. Although beverage consumption habits form during young adult-hood, which has a strong impact on beverage choices in later life, nutrition education programs on beverages are scarce in this population. The purpose of this investigation was 1) to assess soft drink and milk consumption and 2) to evaluate the effectiveness of 15-week class-based nutrition intervention in changing beverage choices among college students.

Methods

A total of 80 college students aged 18 to 24 years who were enrolled in basic nutrition class participated in the study. Three-day dietary records were collected, verified, and analyzed before and after the intervention. Class lectures focused on healthful dietary choices related to prevention of chronic diseases and were combined with interactive hands on activities and dietary feedback.

Results

Class-based nutrition intervention combining traditional lecture and interactive activities was successful in decreasing soft drink consumption. Total milk consumption, specifically fat free milk, increased in females and male students changed milk choice favoring skim milk over low fat milk. (1% and 2%).

Conclusion

Class-based nutrition education focusing on prevention of chronic diseases can be an effective strategy in improving both male and female college students' beverage choices. Using this type of intervention in a general nutrition course may be an effective approach to motivate changes in eating behaviors in a college setting.

Background

In the USA, carbonated soft drinks and milk are the two most popular non-alcoholic beverages, accounting for 39.1% of total beverage consumption [1]. Soft drink consumption has exploded over the past three decades [2] demonstrating a per capita availability increase from 22 gallons to 52 gallons [3,4]. Sugar sweetened soft drinks became a major source of added sugar in the American diet [5,6] and have been linked to adverse nutritional and health consequences such as dental caries and obesity [5,7-12]. Furthermore, evidence also supports an association between soft drink consumption and decreased bone mineral density (BMD) [8,13,14].

Milk and other dairy products are the major source of dietary calcium contributing to about 70% of the calcium in the U.S. food supply [3]. Sixty years ago, Americans drank more than four times more milk as compared to soft drinks, but 2 1/3 times more soft drinks were consumed than milk by 1998 [3]. This trend demonstrates a possible displacement of milk intake [15]. In addition, data showed that between age 6 and 19 years, age is positively associated with soft drink consumption and negatively with milk intake [16]. This relationship is most prevalent in adolescents and young adults [13]. Sufficient intake of calcium, especially during adolescence and young adulthood, is important to maximize peak bone mass (PBM). Failure to achieve PBM increases the incidence of osteoporotic fracture later in life [17].

Young adulthood is a unique period whereby youth obtain independence from their parents. People in this age group are vulnerable to develop unhealthy behaviors [18,19], which will predispose them to chronic diseases later in life [20]. A longitudinal study tracking soft drink intake from early adolescence to later adulthood demonstrated that soft drink consumption from young adulthood remained stable [21]. This data indicates beverage consumption habits formed during young adulthood may have a strong impact on beverage choices in later life. In addition, since milk intake decreases with age after childhood, there is an urgent need for tailored nutrition intervention targeting the young adults to improve their beverage choices.

The purpose of this investigation was two-fold:

1. to assess soft drink and milk consumption;
2. to evaluate the effectiveness of 15-week class-based nutrition intervention in changing beverage choices among college students.

Methods

During spring 2006, ninety healthy college students, between the ages of 18 and 24 years, enrolled in a basic sophomore level nutrition class at a Midwest University participated in the study. This research was approved by the University Institutional Review Board and informed consent was obtained from each participant before enrollment in the project.

The present study used a pre-post test design. Data were collected during the first two weeks and the last week of spring semester in 2006. Body weight was measured in kilograms to the nearest 0.1 kilogram on an electronic scale in light clothing without shoes. Standing height was recorded without shoes on a portable stadiometer to the nearest 0.1 centimeter with mandible plane parallel to the floor. Each subject's BMI was calculated as weight (kg)/height 2(m).

Dietary intake was assessed using 3-day dietary records for two typical weekdays and one weekend day. A variety of tools were used to obtain reliable data. Food models, measuring cups and spoons, household utensils, and tableware were used to illustrate proper portion sizes. Participants were asked to collect and bring all the food labels of products they consumed during data collection period. To obtain the most accurate dietary data, research associates visited local restaurants and campus cafeterias where the majority of participants ate to gain accurate information about ingredients and portion sizes. Foods were purchased if needed. Dietary analysis was performed by the same individual using NutriBase IV Clinical (Cyber Soft Inc, Arizona).

The class met three times per week for 50 minutes per session. Class lectures specifically emphasized 1) the importance of nutrition related to prevention of chronic diseases, 2) increasing consumption of fruits, vegetable and whole grain products, 3) encouraging low fat dairy product consumption, 4) discouraging over reliance on dietary supplements and 5) promoting active lifestyle. In addition to the traditional approach by lectures, video-tape watching and various hands-on activities were integrated. Hands-on activities were designed to enable students to translate lecture materials into real life application. For example, after lectures of lipid and calcium, students assessed their risks for heart disease and osteoporosis, by completing risk assessment forms. These activities helped students identify risk factors and realize that they are not free of chronic disease risks just because they are young or currently disease free. In addition, students completed "Happy Body Log" and listed good things that they did for their body in a daily log. The key of this activity was to start with small behavior changes such as: not eating while watching T.V., reducing portions of single condiments, choosing skim milk over 2% milk. Another approach to encourage dietary behavior change included returning the results of dietary analysis to the students. They were asked to bring their returned results to every class. During lectures, students compared their actual intakes to dietary recommendations (i.e. MyPyramid and Dietary Recommended Intake), which allowed them to realize the strengths and weaknesses of their diet.

Descriptive statistics were presented as means and standard errors. Repeated measures ANOVA with gender as a between-subjects factor and time as a within-subjects factor was used to compare consumption of total soft drink, regular soft drink, diet soft drink, total milk, low fat milk, and fat free milk before and after the intervention. Because there were many more females than males in the class, paired t-tests would be heavily biased toward the females. Therefore, the estimated marginal means obtained from a repeated measures linear model were provided for the pre- and post-test, weighting males and females equally. Estimated marginal means were chosen to represent the total effect since the population of interest includes all college students. Among college students, males and females are represented more evenly than in our sample, so such a summary is justified. Significance tests for the marginal means draw power from the full sample, while the effect size is a compromise between the effect sizes for males and females. As a result, it is possible to have non-significant effects for each gender alone, but significant effects for the marginal means. In addition to the values in the table, total calcium intake and calcium intake from milk at pre- and posttest were calculated. Spearman correlations were calculated to quantify correlations among variables before and after the intervention. Correlations were calculated among change scores. Significance was set a priori at $P \leq 0.05$. All analyses were performed using SPSS for Windows (version 15.5, 2007, SPSS, Chicago, Ill). The result of this study is limited to beverage consumption and the results from other data from the food records have been published elsewhere [22].

Results

Among ninety students enrolled in a sophomore level general nutrition course, 80 students completed the study. Participants were mainly females (87.5%) and white (89.7%). Average BMI of the participants was 26.3 ± 5.63 kg/m2. Average age of the participants was 20.15 ± 1.38 years.

Table 1 summarizes the data and statistical tests on change in beverage consumption as a result of the intervention.

Total soft drink consumption significantly decreased from baseline ($P < 0.05$), although there was insufficient evidence to declare a significant difference for either gender alone. There was marginal evidence that regular soft drink consumption at posttest decreased from the baseline. No change in the consumption of diet soft drink was demonstrated.

For total milk, combining results across genders, no significant change was observed. However, the average change in total milk consumption was significantly increased from baseline ($P < 0.05$) for females but not for males. Whole milk consumption at baseline did not change after the intervention in either gender. Low fat milk consumption decreased significantly ($P < 0.05$) due to a significant change in the males' consumption patterns whereby, there was a significant increase in fat free milk intake after the intervention ($P < 0.01$). This effect was observed to be significant in females ($P < 0.05$) and marginally significant in males.

Total calcium intake at pretest was 813.18 ± 501.48 mg and 858.21 ± 373.11 mg at posttest, respectively. Calcium intake contributed by milk consumption was 156.75 mg at the pretest and 233.0 mg after the intervention.

Table 1. Pre- & Posttest Daily Intake of Beverage by Gender (Means ± Standard Errors, Repeated Measures Analysis)

	Gender (n)	Pretest (fl.oz) Mean (SE)	Posttest (fl.oz) Mean (SE)	p-value
Total soft drink				
	Male (9)	8.53 (3.29)	4.74 (2.27)	0.093
	Female (70)	4.94 (0.84)	3.62 (0.62)	0.100
	Estimated Marginal Mean	6.73 (1.30)	4.18 (0.95)	0.033*
Regular soft drink				
	Male (8)	5.33 (3.36)	2.96 (2.28)	0.145
	Female (70)	2.11 (0.45)	1.28 (0.41)	0.072
	Estimated Marginal Mean	3.72 (0.86)	2.30 (0.72)	0.051
Diet soft drink				
	Male (9)	3.79 (2.31)	1.78 (1.35)	0.346
	Female (70)	2.83 (0.74)	2.30 (0.54)	0.490
	Estimated Marginal Mean	3.31 (1.11)	2.04 (0.79)	0.263
Total milk				
	Male (9)	6.62 (2.32)	6.63 (2.41)	0.997
	Female (70)	4.18 (0.71)	6.23 (0.85)	0.022*
	Estimated Marginal Mean	5.40 (1.07)	6.43 (1.26)	0.433
Whole milk				
	Male (9)	0.00	0.00	1.000
	Female (69)	0.58 (0.31)	0.13 (0.13)	0.149
	Estimated Marginal Mean	0.29 (0.44)	0.07 (0.19)	0.621
Low fat milk				
	Male (9)	6.18 (2.42)	1.70 (1.11)	0.020*
	Female (69)	2.16 (0.52)	2.09 (0.54)	0.918
	Estimated Marginal Mean	4.17 (0.84)	1.90 (0.77)	0.027*
Fat free milk				
	Male (9)	0.44 (1.33)	4.93 (7.75)	0.052
	Female (69)	1.67 (0.51)	3.54 (0.78)	0.026*
	Estimated Marginal Mean	1.06 (0.71)	4.23 (1.18)	0.010*

*demonstrates significant difference P ≤ 0.05

Correlation coefficients between milk and soft drink consumption were not significant at baseline and remained unchanged after the intervention. In addition, changes in consumption for each type of drink were not correlated with each other except for an observed negative correlation between the change in fat free milk intake and the change in low fat milk consumption whereby as fat free milk consumption increased low fat milk consumption decreased (r = -0.317, P < 0.05). In addition, there was a positive correlation between milk consumption and dietary calcium intake (r = 0.578, P < 0.001) at baseline, which further increased after the intervention (r =.689, P < 0.001).

Discussion

The results of this study provided evidence that class-based nutrition education was a viable mechanism to use to help college students make positive changes in soft drink and milk consumption. Previous literature has demonstrated that there have been several studies using college nutrition courses to motivate overall dietary changes [23,24]. Results of this research indicated that nutrition courses increased nutrition knowledge but did not promote dietary changes. On the other hand, a study using a college nutrition science course to prevent weight gain in freshmen revealed that class-based nutrition education may help college students translate nutrition knowledge into dietary changes [25]. Overall, prior research on interventions targeting college students' dietary behaviors suggest a need to develop curriculums targeting specific nutrition behaviors in college students.

After the intervention, overall total soft drink consumption had significantly decreased from baseline. The decrease in total soft drink consumption was mainly due to the reduction in regular soft drink consumption because diet soft drink intake did not decrease as a result of the intervention. The general nutrition class designed to increase the awareness of importance of nutrition in prevention of chronic disease through the combination of traditional lecture with interactive activities may have encouraged the students to reduce soft drink consumption as a part of healthy eating practices. Although it is still debated whether soft drink consumption is associated with increasing obesity rates or decreased milk consumption, it is evident that soft drink consumption has been linked to some negative life style and dietary patterns [26-29]. In a cluster study, Kvaavik et al. found that soft drink consumption could be a marker of unhealthy eating behaviors [16] indicating that reduced intake of soft drink in the current investigation may reflect increased overall diet quality by class-based nutrition intervention.

It should be noted that the amount of soft drinks consumed before the intervention was lower than the results reported by other researchers [30,31] who reported daily soft drink intake of young adults between 11 and 14.4 ounces. There are several reasons to explain this discrepancy. In a study of adolescents, Bere et al. [32] reported that the participants who planned to receive college education showed lower odds of drinking soft drink. Cullen et al. [33] also found that lower parental education was associated with higher consumption of soft drinks. This data perhaps suggests that lower soft drink consumption in the current study may have been due to the higher education level of the participants, college students, compared to the study population, a mixture of both college students and young adults not enrolled in college, used in the previous studies [31].

A second positive finding of this study is that, although total milk consumption did not increase significantly between the genders, females increased their total milk consumption by increasing fat free milk intake while maintaining their low fat milk intake at the same level. Daily calcium intake contributed by milk consumption in females was 156.75 mg at the pretest and 233.0 mg after the intervention. This indicates that only 19% of total calcium intake was coming from milk before intervention and 25% after intervention. This is an encouraging finding because females are at an increased risk to develop osteoporosis in later life if calcium intake is compromised during adolescence and young adulthood. Meanwhile, males switched their milk choices from low fat milk to fat free milk since their total milk consumption did not change, which may demonstrate males may not recognize osteoporosis as an immediate danger due to a broad notion that osteoporosis an "old woman's disease" [28]. It may be that the males chose fat free milk over low fat milk in an attempt to reduce fat intake, which was an important educational component in the classroom lectures and projects. However, it should be noted that, even after the intervention, milk and calcium intake was still much lower than the recommended levels, 3 cups per day by My Pyramid [34] or 1000 mg [35] for both genders at pre- or post-test, although total milk consumption increased in females after the intervention. This finding underscores the necessity of nutrition intervention specifically designed to increase calcium intake in college students.

The positive correlation between dietary calcium and milk intake supports the idea that increasing milk consumption is a desirable way to encourage calcium intake to promote adequate bone health.

Over the last two decades, several researchers have reported that a reduction in milk intake coincides with an increased consumption of soft drinks consumption and hypothesized that soft drink has displaced milk [6,15]. However, in agreement with the previous finding by Storey et al. [36], the current study revealed no association between soft drink consumption and milk intake at either baseline or post test, perhaps suggesting that soft drink consumption did not displace milk consumption in this population. This finding may imply that educating individuals to decrease soft drink consumption is not going to directing increase dairy consumption and that further dairy education needs to be addressed to ensure an adequate consumption of dairy products other than milk.

A limitation of this study is that a convenience sample without a control group was used. Therefore, the study population may not represent traditional college students. In addition, possible confounding factors, such as seasonal variation in beverage consumption, were not controlled for.

In conclusion, class based nutrition education intervention which focused on the prevention of chronic diseases has the potential in college students to reduce soft drink consumption and to increase milk consumption, specifically fat free milk, in female students and to alter milk choice in males from low fat milk to skim milk. Using this type of intervention in a general nutrition course may be an effective approach to motivate changes in eating behaviors in a college setting. Considering gender differences in changes in milk intake, future intervention programs may require different strategies for males emphasizing osteoporosis risk in men and the importance of osteoporosis prevention at earlier stages of life.

Competing Interests

The authors declare that they have no competing interests.

Authors' Contributions

EH designed the study. EH and NC were responsible for data collection. EH and CH conducted data analysis. EH, NC, CH and KL interpreted the analysis contributed to writing and revising the manuscript. All authors read and approved the final manuscript.

References

1. American Beverage Association: What America drinks. [http://improvey-ourhealthwithwater.info/a1/whatamericadrinks.pdf]. 2008.

2. Nielsen SJ, Popkin BM: Changes in Beverage Intake between 1977 and 2001. Am J Prev Med 2004, 27:205–210.

3. Gerrior S, Putnam J, Bente L: Milk and milk products: their importance in the American diet. Food Rev 1998, May-Aug:29–37.

4. Jacobson M: Liquid candy. 2nd edition. Washington, DC: Center for Science in the Public Interest; 2005.

5. Bray GA, Nielsen SJ, Popkin BM: Consumption of high-fructose corn syrup in beverages may play a role in the epidemic of obesity. Am J Clin Nutr 2004, 79:537–543.

6. Gurthrie JF, Morton JF: Food sources of added sweeteners in the diets of Americans. J Am Diet Assoc 2000, 100:43–51.

7. Heller K, Burt BA, Eklund SA: Sugared soda consumption and dental caries in the United States. J Dent Res 2001, 80:1949–1953.

8. Ma D, Jones G: Soft drink and milk consumption, physical activity, bone mass, and upper limb fractures in children: A population-based case-control study. Calcif Tissue Int 2004, 75:286–291.

9. Ludwig DS, Peterson KE, Gortmaker SL: Relation between consumption of sugar-sweetened drinks and childhood obesity: a prospective, observational analysis. Lancet 2001, 357:505–508.

10. Tam CS, Garnett SP, Cowell CT, Campbell K, Gabrera , Baur LA: Soft drink consumption and excess weight gain in Australian school students: results from the Nepean study. Int J Obesity 2006, 30:1091–1093.

11. Schulze MB, Manson JE, Ludwig DS, Colditz GA, Stampfer MJ, Willett WC, Hu FB: Sugar-sweetened beverages, weight Gain, and incidence of Type 2 diabetes in young and middle-aged women. JAMA 2004, 292:927–934.

12. Raben A, Vasilaras TH, Moller AC, Astrup A: Sucrose compared with artificial sweeteners: different effects on ad libitum food intake and body weight after 10 wk of supplementation in overweight subjects. Am J Clin Nutr 2002, 76:721–729.

13. Wyshak G: Teenaged girls, carbonated beverage consumption and bone fractures. Arch Pediatr Adolesc Med 2000, 154:610–613.

14. McGartland C, Robson PJ, Murray L, Cran G, Savage MJ, Watkins D, Rooney M, Boreham C: Carbonated soft drink consumption and bone mineral density in adolescence: the Northern Ireland Young Hearts project. J Bone Miner Res 2003, 18:1563–1569.

15. Rampersaud GC, Bailey LB, Kauwell GP: National survey beverage consumption data for children and adolescents indicate the need to encourage a shift toward more nutritive beverages. J Am Diet Assoc 2003, 103:97–100.

16. Forshee RA, Storey ML: Total beverage consumption and beverage choices among children and adolescents. Int J Food Sci Nutr 2003, 54:297–307.

17. Matkovic V, Kostial K, Simonovic I, Buzina R, Brodarec A, Nordin BEC: Bone status and fracture rates in two regions of Yugoslavia. Am J Clin Nutr 1979, 32:540–549.

18. Huang YL, Song WO, Schemmel RA, Hoerr SM: What do college students eat? Food selection and meal pattern. Nutr Res 1994, 14:1143–1153.

19. Hubert HB, Eaker ED, Garrison RT, Castelli WO: Lifestyle correlates of risk factor change in young adults: an eight-year study of coronary heart disease risk factors in the Framingham offspring. Am J Epidem 1987, 125:812–831.

20. Winkleby MA, Cubbin C: Changing patterns in health behaviors and risk factor related to chronic diseases, 1990-2000. Am J Health Promt 2004, 19:19–27.

21. Kvaavik E, Andersen LF, Klepp K: The stability of soft drinks intake from adolescence to adult age and the association between long-term consumption of soft drinks and lifestyle factors and body weight. Pub Health Nutr 2005, 8:149–157.

22. Ha EJ, Caine-Bish N: Effect of nutrition intervention using a general nutrition course for promoting fruit and vegetable consumption among college students. J Nutr Educ Behav 2009, 41:103–109.

23. Skinner JD: Change in students' dietary behavior during a college nutrition course. J Nutr Educ 1991, 23:72–75.

24. Amstutz MK, Dixon DL: Dietary change resulting from the expanded food and nutrition education program. J Nutr Educ 1986, 18:55–69.

25. Matvienko O, Lewis DS, Schafer E: A college nutrition science course as an intervention to prevent weight gain in female college freshmen. J Nutr Educ 2001, 33:95–101.

26. Pereira MA, Kartashov AI, Ebbeling CB, Van Horn L, Slattery ML, Jacobs DR Jr, Ludwig DS: Fast-food habits, weight gain, and insulin resistance (the CARDIA study): 15-year prospective analysis. Lancet 2005, 365:36–42.

27. Hu FB, Li TY, Colditz GA, Willett WC, Manson JE: Television watching and other sedentary behaviors in relation to risk of obesity and type 2 diabetes mellitus in women. JAMA 2003, 289:1785–1791.

28. Taylor A, Gill T, Phillips P, Leach G: A population perspective of osteoporosis. How common? What impact? How modifiable? Health Prom J Australia 2003, 14:61–65.

29. Forshee RA, Storey ML: Total beverage consumption and beverage choices among children and adolescents. Int J Food Sci Nutr 2003, 54:297–307.

30. West DS, Bursac Z, Quimby D, Preweitt TE, Spatz T, Nash C, Mays G, Eddings K: Self-reported sugar-sweetened beverage intake among college students. Obesity 2006, 14:1825–1831.

31. Storey ML, Forshee RA, Anderson PA: Beverage consumption in the US population. J Am Diet Assoc 2006, 106:1992–2000.

32. Bere E, Glomnes ES, Velde SJ, Klepp K: Determinants of adolescents' soft drink consumption. Pub Health Nutr 2007, 11:49–56.

33. Cullen KW, Ash DM, Warneke C, Moor C: Intake of soft drinks, fruit-flavored beverages, and fruits and vegetables by children in grades 4 through 6. Am J Public Health 2002, 92:1475–1477.

34. United States Department of Agriculture: My pyramid: A steps to a healthier you. [http://www.mypyramid.gov]

35. National Academy of Sciences: Institute of Medicine. [http://www.nal.usda.gov/fnic/DRI//DRI_Calcium/71-145.pdf]. Food and Nutrition Board 1997.

36. Storey ML, Forshee RA, Anderson PA: Associations of adequate intake of calcium with diet, beverage consumption, and demographic characteristics among children and adolescents. J Am Coll Nutr 2004, 23:18–33.

Smoking Prevalence Trends in Indigenous Australians, 1994-2004: A Typical Rather than an Exceptional Epidemic

David P. Thomas

ABSTRACT

Background

In Australia, national smoking prevalence has successfully fallen below 20%, but remains about 50% amongst Indigenous Australians. Australian Indigenous tobacco control is framed by the idea that nothing has worked and a sense of either despondency or the difficulty of the challenge.

Methods

This paper examines the trends in smoking prevalence of Australian Indigenous men and women aged 18 and over in three large national cross-sectional surveys in 1994, 2002 and 2004.

Results

From 1994 to 2004, Indigenous smoking prevalence fell by 5.5% and 3.5% in non-remote and remote men, and by 1.9% in non-remote women. In contrast, Indigenous smoking prevalence rose by 5.7% in remote women from 1994 to 2002, before falling by 0.8% between 2002 and 2004. Male and female Indigenous smoking prevalences in non-remote Australia fell in parallel with those in the total Australian population. The different Indigenous smoking prevalence trends in remote and non-remote Australia can be plausibly explained by the typical characteristics of national tobacco epidemic curves, with remote Indigenous Australia just at an earlier point in the epidemic.

Conclusion

Reducing Indigenous smoking need not be considered exceptionally difficult. Inequities in the distribution of smoking related-deaths and illness may be reduced by increasing the exposure and access of Indigenous Australians, and other disadvantaged groups with high smoking prevalence, to proven tobacco control strategies.

Introduction

Australia has successfully reduced national daily smoking prevalence (aged 14 and over) to 16.6% [1], and national tobacco control policy debate is now concentrated on how to reduce national smoking prevalence to 9% by 2020 [2]. In contrast, the most recent national survey of Indigenous Australians reported that 50% aged 18 and over were daily smokers [3]. There are two distinct groups of Indigenous peoples in Australia: Aboriginal peoples and Torres Strait Islanders. One-quarter of Indigenous Australians (compared with 2% of the total Australian population) live in remote areas with less access to most services. Nevertheless Indigenous Australians live in all the parts of the country, including all states and territories, and comprise 2.5% of the Australian population (about 90% of whom are Aboriginal people) but have much poorer health than other Australians [4]. A public campaign to reduce this health inequality with the slogan 'Close the Gap' was launched in 2007 by Indigenous, human rights and other non-government organizations [5]. The language of the campaign was adopted by the Australian Labor Party, first in Opposition and then in government following the 2007 Australian Federal election.

Reducing Aboriginal and Torres Strait Islander smoking and smoking-related harm is a central element in the Australian government's efforts to improve Indigenous health and to 'Close the Gap' between Indigenous and other Australians.

Smoking is estimated to cause 20% of Indigenous deaths, and to be responsible for 17% of the health gap between Indigenous and other Australians [6,7]. Smoking is the largest single risk factor contributing to the Indigenous disease burden and the health gap with other Australians. In March 2008, Australian Prime Minister Rudd announced that his government would spend an additional $14.5 million over four years on Indigenous tobacco control as one of his government's first steps to 'Close the Gap' [8]. Later that year, further funds for tackling Indigenous smoking were announced as part of the $1.6 billion investment by the Council of Australian Governments in 'Closing the gap' [9].

Whilst Australian tobacco control advocates struggle with reduced media and public interest due to the perception that all their battles have been won [10], Australian Indigenous tobacco control is framed by the idea that nothing has worked and a sense of either despondency or the difficulty of the challenge. Several recent government and research reports comment that the prevalence of smoking in the Australian Indigenous population has not changed, whilst smoking has fallen in the Australian population [4,11-14]. These comments are mainly based on the estimates of total Indigenous smoking prevalence in three national Indigenous surveys performed by the Australian Bureau of Statistics (ABS) in 1994, 2002 and 2004: the National Aboriginal Torres Strait Islander Survey (NATSIS), Social Survey (NATSISS), and Health Survey (NATSIHS). Some, however, also use the smaller Indigenous samples in the 1995 and 2001 National Health Surveys. However, as is only rarely noted [15], the different survey reports used different age cut-offs or different definitions of smoking. Researchers have not performed the detailed examination of Indigenous smoking prevalence trends as has occurred for the total Australian population [16]. A notable exception is the recent paper describing falling smoking prevalence from 1996 to 2005 in both Indigenous and non-Indigenous students aged 12 to 17 years participating in triennial surveys of secondary school students [17].

This paper examines the trends in smoking prevalence of Indigenous men and women aged 18 and over in three large national surveys in 1994, 2002 and 2004, and the implications for tobacco control in this high smoking prevalence disadvantaged group. The next national Indigenous survey has been completed, however, results will not be available until 2010.

Methods

The National Aboriginal Torres Strait Islander Survey was conducted from April to July 1994, the National Aboriginal Torres Strait Islander Social Survey from August 2002 to April 2003, and the National Aboriginal Torres Strait Islander Health Survey from August 2004 to July 2005 [3,18,19]. All three surveys used multi-stage sampling strategies. The 1994 NATSIS sampled all residents from randomly selected households from a stratified sample of Census Collection Districts. The 2002 NATSISS and 2004 NATSIHS randomly sampled up to three residents from randomly selected households from either a stratified sample of Census Collection Districts or a random sample of discrete Indigenous communities and outstations. Non-private dwellings (e.g. hostels, hospitals, caravan parks and prisons) were only sampled in the first survey, but these were excluded from the file analysed in this paper [20]. Response rates were about 80-90% in each survey [3,19,21].

Analyses used STATA Version 10 with the Confidentialised Unit Record Files (CURF) for each survey via ABS's Remote Data Laboratory. Under this arrangement, researchers do not have direct access to the datasets, but instead submit statistical code to ABS, which runs the commands and returns the results to the researchers. All analyses used the expansion factor (or person weight) for each respondent to adjust for the disproportionate sampling of some groups, and so estimates reflect the total Indigenous population not just the sample [22]. These weights are based on the Indigenous estimated resident population in private dwellings on 30 June 1994, 31 December 2002 and 2004 [3,19,20]. Confidence intervals were calculated using the replicate weights for each person generated by ABS [22]. As ABS only created 100 replicate weights for the first survey but 250 for the subsequent surveys, it was not possible to combine the files to directly compare smoking prevalences between surveys (and estimate the confidence intervals of any differences) or to build logistic regression models to examine trends in more detail.

The surveys differed in the ways that smoking status was determined and other characteristics of respondents were classified. The first survey only reported whether people smoked or did not; the second survey asked if current smokers were daily smokers or not; the third survey asked if current smokers were either daily, at least weekly or less than weekly smokers. So even though the reports based on the last two surveys describe prevalences of daily smokers, this paper concentrates on prevalences of current smokers: the only consistent category. The last survey only asked people aged 18 and over about smoking, so this paper concentrates on this age group even though younger people were asked the question in earlier surveys. Variables for state or territory of Australia were not included in the CURF for the first survey, so the analyses using state or territory for this survey were performed separately by ABS using the original dataset. Remote regions include those classified by the Accessibility/Remoteness Index of Australia (ARIA) as remote or very remote and include most of the continent: all of the arid inland and almost all of the tropical North. Indigenous status was only available for those in Queensland in the last two surveys (62% of Torres Strait Islanders lived in Queensland in 2006) [4]; in the last survey if respondents were both Torres Strait Islander and Aboriginal they were classified as Torres Strait Islander, so I treated those in the 2002 survey similarly even though a separate category for both Aboriginal and Torres Strait Islander was available. In the first survey, those who listed both Indigenous groups, were classified as Aboriginal if they listed Aboriginal before Torres Strait Islander, and vice versa.

There were 7,710 and 8,523 and 5,757 Indigenous people aged 18 and over who responded to the 1994, 2002 and 2004 surveys. Data on smoking status was missing for ten, seventy and one respondents aged 18 and over in the three surveys. These non-respondents were excluded in all calculations of smoking prevalence.

Ethical approval was given by the Human Research Ethics Committee of the NT Department of Health and Families and Menzies School of Health Research, including its Aboriginal subcommittee.

Results

Smoking prevalence in the Australian Indigenous population 18 and over declined by 2.4% from 1994 to 2004: from 54.5% (95% CI 51.7-57.4) in 1994 to 53.5(CI 51.0-56.0) in 2002 to 52.1% (CI 49.9-54.3) in 2004.

Figure 1 shows different smoking prevalences and trends for men compared to women and for remote and non-remote Australia (and the corresponding total Australian smoking prevalence trends)[16]. Most differences are small, with overlapping confidence intervals suggesting that differences may have occurred by chance, but some consistent trends are apparent (see Tables 1 and 2). There were much larger differences between the smoking prevalences of men and women in remote regions, which decreased from 1994 to 2004, than between men and women in non-remote regions. In each survey, men in remote regions had the highest smoking prevalence and women in remote regions the lowest; men and women from non-remote regions had similar smoking prevalences between these extremes. From 1994 to 2004, smoking prevalence fell by 5.5% and 3.5% in non-remote and remote men, and by 1.9% in non-remote women. In contrast, smoking prevalence rose by 5.7% in remote women from 1994 to 2002, before falling by 0.8% between 2002 and 2004.

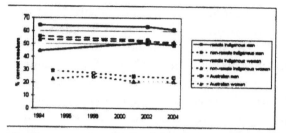

Figure 1. Prevalence of smoking among Australians aged 18 and over, 1994 to 2004-Indigenous men and women in remote and non-remote areas compared with all Australian men, all Australian women. Sources: Weighted data from National Aboriginal Torres Strait Islander Survey 1994, the National Aboriginal Torres Strait Islander Social Survey 2002 and the National Aboriginal Torres Strait Islander Health Survey 2004; and Winstanley and White (2008).

Table 1. Percentages of Indigenous men aged 18 and over who smoked in each survey

	1994	2002	2004
Age (years)			
18-24	58.5 (51.1-65.9)	58.3 (50.5-66.1)	53.6 (47.0-60.3)
25-34	65.5 (58.7-72.2)	58.1 (52.4-63.7)	57.3 (50.3-64.2)
35-44	60.1 (54.0-66.2)	60.5 (54.2-66.8)	59.2 (54.1-64.3)
45-54	52.8 (45.5-60.0)	52.1 (45.1-59.1)	51.5 (43.6-59.5)
55+	42.4 (28.0-56.8)	41.6 (32.8-50.3)	36.2 (28.8-43.5)
Region			
Remote	64.5 (60.9-68.1)	63.3 (60.0-66.6)	61.0 (56.4-65.7)
Non-remote	55.8 (50.8-60.7)	52.8 (48.7-56.8)	50.3 (46.2-54.5)
Jurisdiction			
New South Wales	56.3 (47.5-65.1)	53.0 (45.9-60.1)	52.8 (45.6-60.1)
Victoria	54.4 (40.1-68.7)	52.1 (44.0-60.2)	57.4 (44.1-70.7)
Queensland	57.7 (50.4-65.0)	58.5 (52.3-64.8)	52.7 (46.5-58.9)
South Australia	64.7 (54.9-74.5)	51.5 (44.5-58.5)	59.1 (50.3-67.9)
Western Australia	56.9 (51.4-62.4)	52.2 (45.6-58.9)	43.5 (36.4-50.6)
Northern Territory	66.2 (60.5-71.9)	65.6 (59.9-71.2)	64.4 (57.1-71.8)
Tasmania/ACT	55.8 (46.8-64.8)	50.2 (43.9-56.5)	46.4 (38.7-54.1)
Indigenous status (Queensland only)*			
Torres Strait Islander	48.5 (35.0-62.0)	50.4 (41.7-59.1)	45.9 (35.0-56.8)
Aboriginal	59.6 (51.4-67.8)	59.8 (52.0-67.5)	54.6 (47.6-61.6)

Sources: Weighted data from National Aboriginal Torres Strait Islander Survey 1994, the National Aboriginal Torres Strait Islander Social Survey 2002 and the National Aboriginal Torres Strait Islander Health Survey 2004.
* Torres Strait Islanders in Queensland who also identified as Aboriginal were classified as Torres Strait Islander in 2002 and 2004, but slightly differently in 1994.

Table 2. Percentages of Indigenous women aged 18 and over who smoked in each survey

	1994	2002	2004
Age (years)			
18-24	52.6 (46.5-58.8)	56.9 (50.0-63.8)	51.4 (44.7-58.5)
25-34	60.4 (55.5-65.4)	57.2 (52.8-61.7)	55.0 (49.6-60.3)
35-44	53.2 (46.0-60.4)	55.0 (50.0-60.0)	58.7 (53.2-64.2)
45-54	45.7 (38.2-53.1)	46.4 (39.5-53.3)	52.3 (45.7-58.8)
55+	24.4 (15.5-33.3)	30.2 (23.9-36.5)	26.7 (20.1-33.3)
Region			
Remote	45.0 (40.7-49.4)	50.7 (46.4-55.0)	49.5 (44.8-54.2)
Non-remote	53.5 (49.5-57.4)	51.8 (48.2-55.3)	51.6 (47.9-55.3)
Jurisdiction			
New South Wales	56.3 (49.8-62.8)	54.7 (48.5-60.9)	54.1 (47.1-61.2)
Victoria	65.4 (53.1-77.7)	39.5 (33.4-45.6)	46.9 (35.4-58.4)
Queensland	49.8 (43.9-55.7)	49.5 (44.2-54.8)	49.0 (43.7-54.3)
South Australia	55.9 (45.5-66.3)	48.4 (40.5-56.4)	52.7 (45.8-59.7)
Western Australia	48.9 (41.8-56.0)	48.3 (41.3-55.2)	51.8 (45.5-58.1)
Northern Territory	39.1 (32.8-45.4)	51.5 (44.4-58.5)	47.8 (40.0-55.5)
Tasmania/ACT	44.4 (33.8-55.0)	45.3 (40.3-50.4)	53.1 (46.0-60.1)
Indigenous status (Queensland only)*			
Torres Strait Islander	39.7 (27.4-52.0)	50.5 (41.0-60.1)	45.9 (36.8-55.0)
Aboriginal	52.0 (45.9-58.1)	49.0 (42.7-55.2)	49.8 (43.7-55.9)

Sources: Weighted data from National Aboriginal Torres Strait Islander Survey 1994, the National Aboriginal Torres Strait Islander Social Survey 2002 and the National Aboriginal Torres Strait Islander Health Survey 2004.
* Torres Strait Islanders in Queensland who also identified as Aboriginal were classified as Torres Strait Islander in 2002 and 2004, but slightly differently in 1994.

In New South Wales, the Australian jurisdiction with the largest Indigenous population, there were small consistent falls of 3.5% and 2.2% in smoking prevalence in men and women, from 1994 to 2004. In Western Australia, smoking prevalence in men fell by a large 13.4% from 1994 to 2004. In contrast, smoking prevalence in women in the Northern Territory rose by 12.4% from 1994 to 2002, before falling slightly by 2004. In Queensland, Torres Strait Islander men and women had consistently lower smoking prevalences than Aborigines in the surveys.

Discussion

The lack of comparability between national Indigenous survey results has seriously limited the ability of Australian tobacco control advocates and policy makers to accurately assess progress in reducing Indigenous smoking. By reanalysing the surveys using standardised classifications of smoking status and age, this study goes some way to rectifying this deficiency.

While the insufficient power of the surveys to more precisely measure smoking prevalence, and so identify small changes between surveys, suggests caution in the interpretation of the results, this study indicates that it may not be true that Indigenous smoking prevalences have remained largely unchanged whilst Australian smoking prevalences have fallen, as has conventionally been stated by many in the past. Australian smoking prevalences in men and women aged 18 and over fell by 5% (29 to 24%) and 2% (23 to 21%) from 1995 to 2004 [16]. Almost all (98%) of the non-Indigenous population live in non-remote regions [4]. Male and female Indigenous smoking prevalences in non-remote Australia fell by 5.5% and 1.9% in parallel with these total Australian smoking prevalences, albeit from a much higher initial prevalence in 1994.

Accelerations and decelerations in the decline in Australian smoking prevalence has been noted to be associated with the level of tobacco control advocacy, legislative activity, taxation (and so the price of cigarettes), and national expenditure on social marketing and other tobacco control activities—with most of 1990s being a period of low tobacco control activity and slower falls in smoking prevalence [16,23,24]. It is not possible with only three surveys to make similar claims about the association between the rates of decline in Indigenous smoking prevalences, in either remote or non-remote regions, and the level of total and specifically targeted Indigenous tobacco control activity.

Australian Indigenous smoking prevalences have also not been resiliently static in remote regions, where one quarter of the Indigenous population lives [4]. The declining male smoking prevalence from very high levels and the rise of female smoking prevalence to a lower peak can be explained neatly by the typical characteristics of the stages and shape of the national tobacco epidemics in men and women. Lopez and colleagues describe male smoking prevalence rising and then falling first, with female smoking prevalence rising more slowly, reaching a lower peak then initially falling more slowly than male prevalence [25]. The remote and non-remote Indigenous smoking trends suggest that remote Indigenous Australia is just at an earlier point in the tobacco epidemic than non-remote Indigenous people, plausibly reflecting later access to commercial cigarettes and later and less exposure to tobacco control activities. Sadly, this typical pattern of the tobacco epidemic, and the lag between peaks in smoking prevalence and mortality, predicts that smoking-attributable Indigenous deaths, at the very least amongst remote women, will continue to rise for some years, regardless of any increased tobacco control activities. Many Indigenous premature deaths could have been averted if Indigenous people had been exposed to more intensive tobacco control activities much earlier in the Indigenous smoking epidemic. The reasons for inadequate Indigenous exposure to tobacco control activity may just be the same as the reasons for less Indigenous access to other health services, but may also include the relative neglect of tobacco control compared to other Indigenous health priorities.

The remote and non-remote classifications conflate considerable heterogeneity in smoking prevalence. For example, all of the NT except its capital city Darwin and its immediate environs is classified as remote, yet there is a more than ten-fold difference in lung cancer incidence between its East Arnhem and Alice Springs Rural regions [26], reflecting dramatically different smoking prevalences two decades earlier [27]. All remote (or non-remote) regions are unlikely to have the same smoking prevalence or be at the same point in the tobacco epidemic.

It is difficult to neatly interpret the different Indigenous smoking trends in the different Australian states and territories. Firstly, and most importantly, interpretation is hampered by the smaller subgroup sample sizes and consequently large confidence intervals. Secondly, jurisdictions have different mixes, which cannot be neatly unscrambled, of two factors that could influence Indigenous smoking trends: the proportions of the Indigenous population who live in remote and non-remote areas (and so who are at different stages of the tobacco epidemic) and the amount of generic and targeted Indigenous tobacco control activity.

The main limitation of this study is that almost all differences were not statistically significant; however, some clear patterns emerged. Only by including questions about smoking in the five-yearly Australian Census could these concerns about statistical power be completely addressed. Larger regular national Indigenous surveys are probably impractical: even the smallest of these three surveys interviewed 1 in 45 of the total Australian Indigenous population and took a year to complete interviews [3]. Smokers may have responded differently to the different smoking questions, with their different categories in the three surveys. Some smokers, especially those smoking less than daily, may not have said they were smokers in response to the single question in 1994 [28]. This would mean we have slightly under-estimated the falls in Indigenous smoking.

Nevertheless, it should be possible with consistent smoking questions in new national Indigenous surveys, which are now scheduled to occur every three years, to slowly build an increasingly precise picture of the trends in Australian Indigenous smoking prevalence. More thorough analyses of trends would be possible if ABS provided the data with the same number of replicate weights for each national Indigenous survey so that results could be properly compared and combined. This monitoring should form an essential part of recently accelerated Australian efforts in Indigenous tobacco control [29].

Conclusion

This study has implications for what tobacco control activities need to be included in future efforts to reduce Indigenous smoking. In the past, the apparent immobility of Indigenous smoking prevalence, whilst total Australian smoking prevalence was successfully falling, could be cited as justification to entirely re-think and re-fashion tobacco control for this Indigenous context [30]. Whilst not denying that every population and setting is different and that tobacco control, like other health promotion, should be sensitive to and acknowledge the local context, we should no longer say that Australian Indigenous smoking is so different that we need to abandon all the strategies that have been proven so effective in the rest of the Australian population and elsewhere. There can be space to trial innovative ideas but the emphasis should be on established tobacco control activities, which should be evaluated in the different local Indigenous settings, and which can readily be made consistent with the key principles for Indigenous tobacco control that were proposed during consultations with Indigenous groups [31]. Indigenous tobacco control need not be considered exceptional, nor should reducing Indigenous smoking be considered exceptionally difficult. However, this study should not be cause for any self-congratulation in Indigenous tobacco control: very high Indigenous smoking prevalences have caused too many premature deaths that could have been prevented by additional and better tobacco control activity.

The implications of this study extend to tobacco control in groups with high smoking prevalence in other countries. First, there is the warning to read closely the published series of reports of national omnibus surveys; in different years, reports may use different categories of smoking status or different age cut-offs and may not always focus on the different smoking prevalences of men and women. More importantly, we should not immediately assume that groups with high smoking prevalence are resistant to the tobacco control activities that are known to be most effective. Such groups may be similarly responsive, but just starting from higher smoking prevalences as they are at an earlier stage in the epidemic.

In Australia, mass media led campaigns have led to falls of similar magnitude in the smoking prevalence in disadvantaged groups (with high smoking prevalence) and less disadvantaged socio-economic groups [32]. An international review of the impact of population-based tobacco control activities on social inequalities in smoking, found overall that tobacco control activities have a similar impact on disadvantaged and less disadvantaged socio-economic groups, but that increased tobacco taxation has a greater impact on more disadvantaged groups [33]. Concerns about the possible lesser impact of smokefree environments legislation on the most disadvantaged may lessen as this legislation expands to more of the public and private spaces used by disadvantaged workers and unemployed people. This is an example of increasing the exposure and access of the most disadvantaged groups to proven tobacco control strategies, rather concentrating than re-inventing new strategies for these groups. In New Zealand, Wilson and colleagues have shown how such enhanced tobacco control has the potential to reduce the mortality gap between Māori (the indigenous people of New Zealand) and those of New Zealand European ethnicity [34]. Increasing the exposure of Indigenous Australians to proven tobacco control strategies also has the potential to 'Close the Gap' in Australia.

Competing Interests

The author declares that they have no competing interests.

Acknowledgements

Thanks to Michelle Scollo and Steve Guthridge for useful comments on an earlier draft. This research is funded by the Cooperative Research Centre for Aboriginal Health.

References

1. Australian Institute of Health and Welfare: 2007 National Drug Strategy Household Survey: detailed findings. Drug Statistics Series no. 22. Cat. No. PHE 107. Canberra: Australian Institute of Health and Welfare; 2008.

2. Tobacco Working Group: National Preventative Health Taskforce. Technical Report No 2. Tobacco control in Australia: Making smoking history. Canberra: Australian Government, National Preventative Health Taskforce; 2008.

3. Australian Bureau of Statistics: National Aboriginal and Torres Strait Islander Health Survey, Australia 2004–05. Cat. No. 4715.0. Canberra: Australian Bureau of Statistics; 2006.

4. Australian Bureau of Statistics, Australian Institute of Health and Welfare: The health and welfare of Australia's Aboriginal and Torres Strait Islander peoples 2008. Cat. No. 4704.0. Canberra: Australian Bureau of Statistics; 2008.

5. Close the gap. National Indigenous Health Equality Targets. Outcomes from the National Indigenous Health Equality Summit, Canberra, March 18–20, 2008, [http://www.hreoc.gov.au/social_Justice/health/targets/index.html].

6. Vos T, Barker B, Stanley L, Lopez AD: The burden of disease and injury in Aboriginal and Torres Strait Islander peoples 2003. Brisbane: School of Population Health, University of Queensland; 2007.

7. Vos T, Barker B, Stanley L, Lopez AD: The burden of disease and injury in Aboriginal and Torres Strait Islander peoples 2003: Policy Brief. Brisbane: School of Population Health, University of Queensland; 2007.

8. Joint Media Release with the Minister for Health and Ageing and the Minister for Indigenous Affairs—Rudd Government Tackles Indigenous Smoking Rates and Health Workforce in next Down Payments on Closing the Gap. 20 March 2008, [http://www.pm.gov.au/node/5890].

9. Joint Media Release—$1.6 billion COAG investment in Closing the Gap. 30 November 2008, [http:/ / www.health.gov.au/ internet/ ministers/ publishing. nsf/ Content/ mr-yr08-nr-nr164.htm].

10. Chapman S: Introduction. [http://www.tobaccoinaustralia.org.au] In Tobacco in Australia: Facts and issues Third edition. Edited by: Scollo MM, Winstanley MH. Melbourne: Cancer Council Victoria; 2008.

11. Australian Institute of Health and Welfare: Aboriginal and Torres Strait Islander Health Performance Framework, 2008 Report: Detailed analyses. Cat. No. IHW 22. Canberra: AIHW; 2008.

12. Winstanley M: Tobacco use among Aboriginal and Torres Strait Islanders. [http://www.tobaccoinaustralia.org.au] In Tobacco in Australia: Facts and issues Third edition. Edited by: Scollo MM, Winstanley MH. Melbourne: Cancer Council Victoria; 2008.

13. Thomas DP, Briggs V, Anderson IPS, Cunningham J: The social determinants of being an Indigenous non-smoker. Aust N Z J Public Health 2008, 32:110–116.

14. Wood L, France K, Hunt K, Eades S, Slack-Smith L: Indigenous women and smoking during pregnancy: Knowledge, cultural contexts and barriers to cessation. Soc Sci Med 2008, 66:2378–2389.

15. Smoking in the Northern Territory. Health Gains Planning fact sheet 2 Darwin: Department of Health and Community Services, Northern Territory Government; 2006.

16. Winstanley M, White V: Trends in the prevalence of smoking. [http://www.tobaccoinaustralia.org.au] In Tobacco in Australia: Facts and issues Third edition. Edited by: Scollo MM, Winstanley MH. Melbourne: Cancer Council Victoria; 2008.

17. White V, Mason T, Briggs V: How do trends in smoking prevalence among Indigenous and non-Indigenous Australian secondary students between 1996 and 2005 compare? Aust N Z J Public Health 2009, 33:147–153.

18. Australian Bureau of Statistics: National Aboriginal and Torres Strait Islander Survey 1994: Detailed Findings. Cat. No. 4190.0. Canberra: Australian Bureau of Statistics; 1996.

19. Australian Bureau of Statistics: National Aboriginal and Torres Strait Islander Social Survey 2002. Cat. No. 4714.0. Canberra: Australian Bureau of Statistics; 2004.

20. Australian Bureau of Statistics: National Aboriginal and Torres Strait Islander Survey: Confidentialised Unit Record File, Technical Manual. Australia 1994 (Reweighted). Cat. No. 4188.0.55.003. Canberra: Australian Bureau of Statistics; 2006.

21. Australian Bureau of Statistics: National Aboriginal and Torres Strait Islander Survey: An evaluation of the survey. Cat. No. 4184.0. Canberra: Australian Bureau of Statistics; 1996.

22. Donath SM: How to calculate standard errors for population estimates based on Australian National Health Survey data. Aust N Z J Public Health 2005, 29:565–571.

23. Hill DJ, White VM, Scollo MM: Smoking behaviours of Australian adults in 1995: trends and concerns. Med J Aust 1998, 168:209–213.

24. White V, Hill D, Siahpush M, Bobevski I: How has the prevalence of smoking changed among Australian adults? Trends in smoking prevalence between 1980 and 2001. Tob Control 2003, 12(Suppl II):ii67–ii74.

25. Lopez AD, Collishaw NE, Piha T: A descriptive model of the cigarette epidemic in developed countries. Tob Control 1994, 3:242–247.

26. Condon JC, Zhang X, Li SQ, Garling LS: Northern Territory Cancer Incidence and Mortality by Region, 1991–2003. Darwin: Department of Health and Community Services; 2007.

27. Watson C, Fleming J, Alexander K: A survey of drug use patterns in Northern Territory Aboriginal communities: 1986–1987. Darwin: Northern Territory Department of Health and Community Services, Drug and Alcohol Bureau; 1988.

28. Mullins R, Borland R: Changing the way smoking is measured among Australian adults:A preliminary investigation of Victorian data. [http://www.quit.org.au/downloads/QE/QE9/QE9Home.html] In Quit Evaluation Studies No9, 1996–1997 Edited by: Trotter L, Mullins R. Melbourne: Victorian Smoking and Health Program; 1998.

29. World Health Organization: WHO Report on the Global Tobacco Epidemic, 2008: The MPOWER package. Geneva: WHO; 2008.

30. Centre for Excellence in Indigenous Tobacco Control (CEITC): Indigenous tobacco control in Australia: Everybody's business. National Indigenous tobacco control research roundtable report, Brisbane, Australia, 23 May 2008. Melbourne: CEITC, University of Melbourne; 2008.

31. Briggs VL, Lindorff KJ, Ivers RG: Aboriginal and Torres Strait Islander Australians and tobacco. Tob Control 2003, 12(Suppl 2):ii5–8.

32. Scollo M, Siahpush M: Smoking and social disadvantage. [http://www.tobaccoinaustralia.org.au] In Tobacco in Australia: Facts and issues Third edition. Edited by: Scollo MM, Winstanley MH. Melbourne: Cancer Council Victoria; 2008.

33. Thomas S, Fayter D, Misso K, Ogilvie D, Petticrew M, Snowden A, Whitehead M, Worthy G: Population tobacco control interventions and their effects on social inequalities in smoking: systematic review. Tob Control 2008, 17:230–237.

34. Wilson N, Blakely T, Tobias M: What potential has tobacco control for reducing health inequalities? The New Zealand situation. Int J Equity Health 2006, 5:14.

Collecting Household Water Usage Data: Telephone Questionnaire or Diary?

Joanne E. O'Toole, Martha I. Sinclair and Karin Leder

ABSTRACT

Background

Quantitative Microbial Risk Assessment (QMRA), a modelling approach, is used to assess health risks. Inputs into the QMRA process include data that characterise the intensity, frequency and duration of exposure to risk(s). Data gaps for water exposure assessment include the duration and frequency of urban non-potable (non-drinking) water use. The primary objective of this study was to compare household water usage results obtained using two data collection tools, a computer assisted telephone interview (CATI) and a 7-day water activity diary, in order to assess the effect of different methodological survey approaches on derived exposure estimates. Costs and logistical aspects of each data collection tool were also examined.

Methods

A total of 232 households in an Australian dual reticulation scheme (where households are supplied with two grades of water through separate pipe networks) were surveyed about their water usage using both a CATI and a 7-day diary. Householders were questioned about their use of recycled water for toilet flushing, garden watering and other outdoor activities. Householders were also questioned about their water use in the laundry. Agreement between reported CATI and diary water usage responses was assessed.

Results

Results of this study showed that the level of agreement between CATI and diary responses was greater for more frequent water-related activities except toilet flushing and for those activities where standard durations or settings were employed. In addition, this study showed that the unit cost of diary administration was greater than for the CATI, excluding consideration of the initial selection and recruitment steps.

Conclusion

This study showed that it is possible to successfully 'remotely' coordinate diary completion providing that adequate instructions are given and that diary recording forms are well designed. In addition, good diary return rates can be achieved using a monetary incentive and the diary format allows for collective recording, rather than an individual's estimation, of household water usage. Accordingly, there is merit in further exploring the use of diaries for collection of water usage information either in combination with a mail out for recruitment, or potentially in the future with Internet-based recruitment (as household Internet uptake increases).

Background

A formal risk management process is increasingly being employed in the management of drinking water supplies and has been adopted in drinking water guidelines [1,2]. It has also been used in the design of Australian guidelines for water recycling for non-potable uses [3]. From a health perspective, these guidelines principally focus on microbial hazards and use a Quantitative Microbial Risk Management (QMRA) process for guideline setting. Exposure assessment, one element in QMRA, describes the conditions conducive to human exposure to the risk and typically includes a description of the intensity, frequency and duration of exposure as well as the exposure routes and the people exposed [4]. Data gaps for water exposure assessment include the duration and frequency of urban non-potable (non-drinking) water use. Good quality contemporaneous data about the duration and frequency of water-using activities (particularly in circumstances where water restrictions may be implemented due to water shortages, as is currently the case in many parts of Australia) combined with information about the total volume per exposure event are needed to obtain the total volume exposure per person per annum. This information, in combination with the number of residual micro-organisms present in the source water and the dose response of micro-organism(s), can then be used to obtain an estimate of the annual probability of infection associated with micro-organism(s) of concern for designated water-using scenarios. Knowing the exposure volume per person per annum for a particular water-using scenario (e.g. toilet flushing) and water type, it is possible to determine the minimum level of water treatment necessary to achieve a predetermined health target and/or to determine whether substitution of one water type with another will lead to an unacceptable increase in the magnitude of risk.

The need for exposure information leads to questions about how such data are best collected. Each survey method has advantages and disadvantages associated with its use. The rationale for use of a particular survey method is often based on its practicality, cost and the complexity of the questions to be answered. This study was undertaken as a sub-set of a larger project that collected information about the duration and frequency of household recycled and drinking water use. The objective of this study was to compare household water usage results obtained using two data collection tools, a computer assisted telephone interview (CATI) and a 7-day water activity diary, in order to assess the effect of different methodological survey approaches on results. This information is important when considering the conduct of future studies, the interpretation of results of any household water usage survey and ultimately, in determining the likely precision of derived exposure estimates. Only 'direct' exposure of householders to recycled water was assessed, with no consideration given to incidental simultaneous exposure of others. Costs and logistical aspects of each survey method were also examined.

Methods

Data were collected from households in the Rouse Hill dual reticulation scheme in Sydney Australia. The Rouse Hill scheme is the largest (approximately 16,000 households) and longest established (since 2001) dual reticulation scheme in Australia. In such schemes households are supplied with two grades of water through separate pipe networks. One grade of water is of high quality and used for drinking, cooking and other household purposes, while the other is of lower quality and used for non-potable purposes such as toilet flushing and garden watering. Whilst water restrictions were applicable to the drinking water supply during the survey period, there was no restriction on the amount of recycled water used by Rouse Hill households. Householders were questioned about their use of recycled water for toilet flushing, garden watering and other outdoor activities. In addition, householders were questioned about their water use in the laundry (currently supplied with drinking water but also a use for which recycled water may be substituted in future). An extensive survey of water-using behaviours had not been administered to Rouse Hill dual reticulation residents before this survey. The study was conducted as a University survey over the period February to April 2006 (inclusive) and was approved by the Monash University Standing Committee on Ethics and Research involving Humans (SCERH Project number 2005/659).

Household Recruitment

Australian Electoral Commission (AEC) and Electronic White Pages (EWP) records were used to select eligible dual reticulation households. In Australia, voting is compulsory and the electoral roll provides an easily accessible and up to date means of contacting persons for health research studies. Records were grouped according to electorate, suburb, postcode, street number and address, resulting in a list of households (single dwellings) located in the area of interest (defined by two postcode zones). A random subset of 3,500 households was selected from the 14,000 available. Data matching with the Electronic White Pages (EWP) was then performed for the purpose of obtaining telephone numbers. The generated random sequence of households was the order in which data matching with the EWP was performed and introductory letters were sent. Telephone matching of records for a total number of 3,500 households only was attempted as a pilot study results indicated that this would be sufficient to achieve a minimum total sample size of 500 CATI responses with 200 of these households also completing and returning a water-activity diary. These sample sizes were selected based on budgetary considerations and to allow a comparison of diary and CATI responses corresponding to a 95% confidence interval of 0.5 - 0.7 where the weighted kappa statistic is 0.6 [5].

Elector households with a listed telephone number in the EWP were sent an introductory letter inviting them to participate in the study. Telephone contact was commenced one to three weeks after the introductory letter was mailed. Four telephone contact attempts for each household were made before attempts were terminated. The majority of telephone calls were made between 6 pm and 9 pm. Once telephone contact was made, householders were invited to complete a Computer Assisted Telephone Interview (CATI). Where this was declined this was the final contact with householders. Households completing the CATI were invited to receive and complete a water-activity diary. Contact was terminated at this stage for those declining to complete a water activity diary. Those agreeing to complete a water-activity diary were sent a set of diary cards and diary instructions and were advised that a gift voucher to the value of A$40 would be posted to them following diary completion and return of the cards.

Computer Assisted Telephone Interview

The CATI questions used for the survey [see additional file 1] were based on a previous Australian household water activity survey [6].

As recycled water in Rouse Hill dual reticulation households is used for toilet flushing and outdoor water use, the CATI was designed to include questions about these activities. Whilst recycled water is not plumbed into Rouse Hill dual reticulation households for machine washing (drinking water is used for this purpose), the CATI was also designed to include questions relating to laundry activities as a scoping exercise to explore the use of recycled water, rather than drinking water, for laundry purposes. The questions posed to householders about the duration of the water-using activities required them to estimate the duration in minutes. For questions about the frequency of a water-using activity, householders were required to classify the frequency of their water usage into pre-defined categories, or in some instances (e.g. toilet flushing, laundry loads and garden watering) to give a number estimate for the 7-days immediately prior to the interview. Since the 1980's, as a water saving measure, Australian regulations have mandated that dual flush toilets are installed in new homes and when existing homes are renovated. Accordingly, respondents were asked to give a number estimate for both half and full toilet flushes per day. Respondents were also required to indicate the machine washing settings (water level, water temperature) commonly used in the household. A total of 523 households completed the CATI.

Water Activity Diary

The water activity diary took the form of 'diary cards' [see additional file 2]. Five different diary cards were produced, each one relating to a particular activity as follows: Household characteristics, toilet use, garden watering, outside water use (excluding garden watering) and laundry use. Each card provided for 7-days recording. The recording of activities required one or more of the following:

- A tick to indicate the frequency of use (e.g. number of toilet flushes (half and full flushes), number of machine washing loads, washing machine settings)
- Circling an option (e.g. washing machine type)

- Entry of a number (e.g. the number of minutes that an activity had been performed)

- The entry of a code (e.g. specification of WCB for watering can or bucket in the outdoor usage diary card with codes specified at the bottom of the card)

- Free-hand written specification of the type of water activity (e.g. for the outdoor usage card where the activity was not already specified on the card) or location of toilet (e.g. ensuite bathroom)

A diary instruction pamphlet was prepared and for clarity of instructions, diary cards were identified by both type and colour. In addition, examples of diary entries for each card were pre-printed on the reverse of each diary card. The diary instructions referred to these examples and provided further explanation. To facilitate the recording of activities as they occurred, it was suggested that the diary cards be placed at convenient locations in the house that made them accessible for recording purposes. For example, it was suggested that the toilet card might be placed on each toilet door. Water proof covers, pens with which to complete diaries and adhesive were provided in each diary pack to support diary completion. In addition, a reply-paid envelope was included with the diary pack to support return of completed diaries. A follow up call was made to householders a few days after mailing the diary pack to check that it had been received, and householders were requested to complete and return the diaries within 4 weeks. There was no further contact with households during the diary recording period.

Data Analysis

CATI responses were entered into an ACCESS (Microsoft Office 2000) database at the time of the telephone interview. Information from each of the diary cards was entered onto an ACCESS (Microsoft Office 2000) diary database which was constructed so that the appearance of the diary cards and the database entry screens were similar. The entry of data from toilet flushing and laundry cards required some minor processing. For the toilet and laundry card this consisted of summing the number of ticks on each of the cards to arrive at tallies for each activity per week. At least 10% and up to 25% of data entry, depending on card type, was checked by a second operator.

Diary numerical entries of the frequency of a water-using activity were classified into the same categories as used for the CATI where there was discordance in data type between the CATI (categorical) and diary (numerical, continuous). For toilet flushing, the number of half and full toilet flushes per household were tallied from diary cards to give a total number of half and full flushes per household per week. For the CATI the number of full and half flushes per day, as estimated by the respondent, was converted to total number of full and half flushes per household per week assuming the same daily toilet flushing estimate for all household members.

Agreement between reported water usage using the CATI and diary responses was measured using a weighted kappa statistic, which considers disagreement close to the diagonal less heavily than disagreement further from the diagonal. Weighted kappa results were classified as 0.00 - 0.40 = 'poor'; 0.41 - 0.75 = 'Fair to good' and 0.76 - 1.00 = 'Excellent' [7]. Bland-Altman scatter plots were used to visually compare the measures of water usage [8]. Statistical analysis was carried out using STATA version 9 (STATA™ Stata Corporation, Texas, USA).

Results

Recruitment

Of the 523 dual reticulation eligible households completing a CATI, 371 (70%) agreed to complete a water activity diary. All 371 households agreeing to complete a diary were sent water activity diary cards and 232 (63%) households returned completed cards within a 4 week period and were eligible for the A$40 voucher offered as an incentive.

Timing of the CATI and Diaries

Administration of CATIs and diaries with the exception of two CATI and seven diaries was compacted into two overlapping periods each of seven weeks duration. There was a two week lag between the administration of the first batch of CATIs and commencement of the first batch of diaries. All diaries except one were commenced within 3 weeks of the cessation of the telephone interviews. The total survey period was 12 weeks.

Comparison of CATI and Diary Characteristics and Time Input

A summary of CATI and diary characteristics, costs, data handling procedures and estimated number of hours for completion of each are given in Table 1.

Table 1. Summary of CATI and diary characteristics and data handling prior to statistical analysis

Characteristic	CATI	Diary
Database design A$	$5000	$2360
Duration	15-20 min*	Not applicable
Target number	500	200
Order of administration	First	Second (always Post CATI)
Recruitment strategy	Electoral records and introductory letter	$40 incentive (voucher)
Time period under investigation	Immediate past week (in detail) and up to one year prior	7-day period only
Data entry onto ACCESS database	Immediate (as telephone interview was conducted)	Manual entry
Actions prior to interview/diary administration excluding initial mail-out and associated activities	- Questionnaire design - ACCESS database design	- Print diary cards - Design and print diary instructions - Compile diary cards; pack and send* - Follow-up phone call to check diary pack receipt* - ACCESS database design
Actions subsequent to interview/diary completion	- Check data entry immediately following interview*	- Check diary cards for completion/legibility* - Tally number of uses recorded on cards* - Transfer data including tallies from diary card to ACCESS database* - Check 10-25% data transferral to database*
Estimated number of hours labour per completed CATI/diary (for * activities)	30 min	1 hr 25 min

Outdoor Water Use

Table 2 gives a summary of CATI and diary responses for the garden irrigation module. These results show lower CATI results for: number of garden watering sessions in prior seven days, duration of automatic water system use and duration of hand held hose use. Higher CATI results were obtained for: duration of use of fixed manual systems and hose and sprinkler watering sessions. A comparison of CATI and diary responses showed poor agreement between data collection tools for all but the duration of the automatic watering system, which showed 'fair to good' agreement (weighted kappa = 0.51). Agreement between data collection tools for the total number of garden watering sessions in the 7-days prior, as measured by a weighted kappa (0.36), was 'poor.' The Bland-Altman scatter plot (Figure 1) for garden watering frequency in each of the 7-day CATI and diary survey periods shows that the variance increased in proportion to the average values. The average difference between the Diary and CATI responses (the diary was higher) was 1.18 (Confidence interval 0.89 to 1.47).

Table 2. CATI versus Diary responses: Garden irrigation

Question	CATI response N = 219		Diary response N = 219		CATI response relative to Diary weighted kappa (agreement[a])
	Average (standard deviation)	% using (N)	Average (standard deviation)	% using (N)	
Number of garden watering sessions last 7 days	1.7 (1.79)	72.2% (158)	2.9 (2.24)	86.3% (189)	0.36 (poor)
Automatic watering system duration use (min)	38.0 (19.2)	15.5% (34)	55.4 (26.4)	11.9% (26)	0.51 (fair to good)
Hand held hose duration use (min)	26.3 (16.9)	40.6% (89)	26.5 (32.7)	58.0% (127)	0.36 (poor)
Fixed manual system duration use (min)	45.3 (23.9)	17.4% (38)	43.2 (17.12)	12.8% (28)	0.21 (poor)
Hose and sprinkler duration use (min)	60.2 (24.8)	13.2% (29)	54.6 (33.5)	22.8% (50)	0.05 (poor)

[a] Fleiss JL: Statistical methods for rates and proportions 2nd edition, New York; Wiley; 1981:218

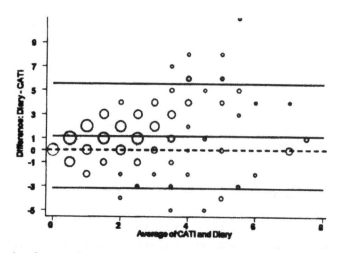

Figure 1. Number of times garden watered in past 7 days. Scatter plot (Bland-Altman plot) of diary response minus CATI response (vertical axis) against average of CATI response and diary response (horizontal axis). The centre full horizontal line corresponds to the average difference and upper and lower full horizontal lines correspond to the 95% limits of agreement. The dashed straight line represents the line of zero difference. The larger the data points (circles) the higher the number of households that recorded these values.

Laundry Use

Agreement between the CATI and diary responses for the number of washing machine loads per week was 'fair to good' as measured by a weighted kappa of 0.66 (Table 3). The Bland-Altman scatter plot (Figure 2) for number of laundry loads per week is symmetric for all but low average values. The average difference (the diary was higher) in the number of machine washing loads per week between the diary and CATI was 1.37 (Confidence interval 1.02 to 1.73). Statistical analysis of the agreement between responses relating to washing machine water level showed 'poor' agreement between data collection tools relating to the selection of low (weighted kappa = 0.24) and medium (weighted kappa = 0.33) water levels and 'fair to good' agreement for high (weighted kappa = 0.44) and automatic (weighted kappa = 0.64) water level settings. Statistical analysis of agreement between CATI responses and the diary relating to water temperature selection showed 'fair to good' agreement relating to selection of cold wash (weighted kappa = 0.62) and warm wash (weighted kappa = 0.58) but only 'poor' agreement for hot wash (weighted kappa = 0.17).

Table 3. CATI versus Diary responses: Laundry

Statistic	Number of washing machine loads per week		CATI response relative to Diary weighted kappa (agreement[a])
	CATI response (Median 95% confidence limit)	Diary response (Median 95% confidence limit)	
Average	5.3	6.7	
Standard deviation	3.1	3.7	
Median	5	6	0.66
	(4 - 5)	(5.5 - 7)	(fair to good)
N	217	217	

[a] Fleiss JL: Statistical methods for rates and proportions 2nd edition, New York; Wiley; 1981:218

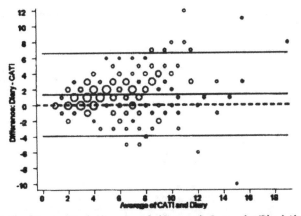

Figure 2. Number of washing machine loads per household per week. Scatter plot (Bland-Altman plot) of diary response minus CATI response (vertical axis) against average of CATI response and diary response (horizontal axis). The centre full horizontal line corresponds to the average difference and upper and lower full horizontal lines correspond to the 95% limits of agreement. The dashed straight line represents the line of zero difference. The larger the data points (circles) the higher the number of households that recorded these values.

Toilet Flushing

Agreement between data collection tools for the total number of toilet flushes per household per week, as measured by a weighted kappa (0.29), was 'poor' (Table 4). The Bland-Altman scatter plot (Figure 3) shows that there was a tendency for greater (negative) differences (diary response lower than the CATI response) with greater average number of toilet flushes per household per week. For the total number of toilet flushes a mean difference between the Diary and CATI estimates of -20.6 (Confidence interval -28.9 to -12.3) was recorded. Table 5 presents estimates of the frequency of use of half flush as compared with full flush by the CATI and diary. A weighted kappa statistic of 0.36 (poor) was obtained when CATI and diary responses were compared for half flush frequency estimates.

Table 4. CATI versus Diary responses: Toilet flushing

Statistic	Total number of toilet flushes per household per week CATI response (Median 95% confidence limit)	Diary response (Median 95% confidence limit)	CATI response relative to Diary weighted kappa (agreement*)
Average	112.4	91.8	
Standard deviation	60.5	40.1	
Median	105	85	0.29
	(96 - 112)	(77 - 90)	(poor)
N	216	216	

* Fleiss JL: Statistical methods for rates and proportions 2nd edition, New York: Wiley; 1981:218

Table 5. CATI versus Diary responses: Toilet flushing

Diary Proportion of time use half flush	CATI: Proportion of time personally use half flush = <25%	25-50%	51-75%	= >75%	Total
= <25%	5	2	2	3	12
25-50%	5	0	6	9	20
51-75%	7	13	25	65	110
= > 75%	0	4	14	61	79
Total	17	19	47	138	221

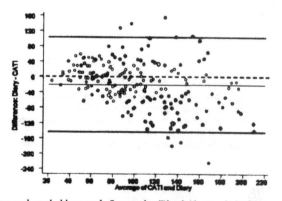

Figure 3. Toilet flushes per household per week. Scatter plot (Bland-Altman plot) of diary response minus CATI response (vertical axis) against average of CATI response and diary response (horizontal axis). The centre full horizontal line corresponds to the average difference and upper and lower full horizontal lines correspond to the 95% limits of agreement. The dashed straight line represents the line of zero difference. The larger the data points (circles) the higher the number of households that recorded these values.

Discussion

When using QMRA techniques to aid in the development of health-based quality targets for waterborne pathogens it is important that input data have validity. One of the inputs required for QMRA is accurate exposure information, which raises questions about how this information is best collected. Validity of a survey method can be quantified by comparing survey responses obtained with a gold standard. For example, in health-related studies where exposure to a particular agent is being assessed, responses to questions about exposures may be compared with results of tests for relevant biological markers. For studies such as this one where water used by householders inside and outside the home for non-potable purposes was the subject of investigation, the true gold standard is the volume of water householders are exposed to during specific water-using activities. Whilst sophisticated water meters are available that can, in theory, measure the volume of in-house and outside house water usage at individual taps, their use is somewhat problematic in an extensive household water usage survey. This is because of cost and logistical considerations associated with meter installation to a large number of households, meter calibration (e.g. to match water usage events such as toilet flushing or showering with recorded data 'spikes') and the lack of sensitivity of meters to detect single tap usage [9]. In addition, the volume of water used at a particular tap is not a true measure of exposure because factors such as human behaviour and the type of water-using equipment also determine the exposure volume.

As a consequence of the barriers associated with accurately measuring individual domestic water exposure, this study sought to collect water usage information using two alternative data collection tools, a telephone interview and water-activity diary, both previously used for Australian household surveys [6,9,10] and compare results. In this study, the comparison of these data collection tools was based not only on estimates of the duration and frequency of water-using activities but also on the cost and logistics of administration of each survey type.

Exposure Estimates

The highest level of agreement between CATI and diary responses was obtained for: number of washing machine loads per week (Table 3); automatic system watering session duration (Table 2); use of high and automatic washing machine levels and use of cold and warm washing machine wash temperatures. Lower levels of agreement between CATI and 7-day diary results were obtained for less frequently performed water-using activities with the exception of toilet flushing (frequently performed but also showing poor agreement) or where standard durations (e.g.

as might occur with automatic pre - programmed systems) or frequently applied settings were not used. 'Poor' agreement was obtained for: number of garden watering sessions; hand held hose watering duration; fixed manual water system duration; hose and sprinkler watering duration (Table 2); washing machine hot water wash; washing machine low water level and washing machine medium water level. Overall, when considering all individual water uses and the relative performance of the diary and CATI, diary responses were not consistently higher or lower than CATI responses.

The poor agreement between diary and CATI responses for the total number of toilet flushes per household per week (weighted kappa = 0.29) (Table 4) and for the frequency of use of half flush as a percentage of total toilet flushes (weighted kappa = 0.36) (Table 5) may be attributed to a number of factors. Considering that the number of toilet flushes is not insignificant per week per individual, it is likely that the exact number at home is difficult to recall. In addition, the CATI respondent was asked about individual behaviour, yet the diary recorded household behaviour. Accordingly, the number of toilet flushes per week estimated by the CATI respondent was multiplied by the number of persons in the households and compared with the tallied entries for toilet flushing recorded in the diary. A comparison of diary and CATI results (Figure 3) showed differential bias namely, that the CATI estimate (single respondent) combined with an assumption of identical toilet flushing behaviour by all persons in the household, overestimates the total flushes in large households.

The higher estimates of the frequency of use of half flush as a proportion of all toilet flushing for the CATI as compared with the diary (Table 5) indicates that the CATI respondent believed that the use of half flush was more prevalent in the household than was the case for the 7-day diary recording period. The CATI responses may represent the desired target behaviour of the respondent rather than actual household practice or, may truly reflect the respondent's behaviour, which is different to the rest of the household.

Many of the activities showing 'poor' agreement between data collection tools, as measured by the weighted kappa statistic, were not only less frequently performed activities, but also outdoor water-related activities. A possible reason for poor agreement between results for outdoor activities is that climatic conditions may have been different for each of the CATI and diary 7-day survey periods. Whilst the majority of diaries and CATIs administered were each compacted into separate seven week periods with no greater than a 3-week lag period between cessation of telephone interviews and commencement of diaries, the climatic conditions during diary and CATI survey periods may have been significantly different for a proportion of households (e.g. rain during the CATI 7-day survey period and no rain during the diary 7-day survey period or vice-versa).

The observation of greater concordance between CATI and diary responses for more frequent and standard water-related activities except toilet flushing may reflect the greater likelihood that standard frequencies, durations and settings are known to all adult household members. This observation is supported by results from other water-related studies where completion of a questionnaire also preceded the diary and referenced a different 7-day period [11,12]. These studies attributed lower correlations between questionnaire and diary responses to a greater potential for daily variation in some water-related activities compared with established routine ones.

When all results of this study are considered, they show that different estimates of the duration and frequency of water-related activities are obtained depending upon the survey method used. For estimates of garden watering frequency and the number of laundry machine loads per week, the diary, compared with the CATI, showed general positive bias (i.e. the diary response was consistently higher). Such differences between responses for CATI and diary may be associated with one or more factors including: recall bias of the CATI respondent [13]; natural variation in water-related activity that occurs from week to week (CATI and diary responses 'referenced' different 7-day periods); the survey period may not have been over a sufficient period to 'capture' the exposure of interest (e.g. if garden watering was performed every 1.5 weeks it may not have been included in either (or both) of the diary or CATI survey periods); computations to convert single respondent results to household results (not all household members behave in the same way) and failure to complete diaries prospectively as intended or to record all activities and events [14].

Compared with a telephone interview, the use of a household diary has a clear advantage in determining household exposure to a particular water supply. This is because the diary provides a collective measure of household exposure compared with an individual's estimate of household exposure. Acknowledging that the household diary cards may have in some instances been completed by one individual, it is more likely that they were completed by the householder performing the activity(ies) of interest. This assumption is based on the cards being placed in the location of water use (if instructions were followed) and/or that responsibility for diary card recording is likely to have been allocated within the household, to the household member with greatest familiarity with household water-using practices. In contrast, the telephone questionnaire may have been answered by an adult not fully familiar with the household water usage.

The prospective recording of water-related activities using a diary is also another advantage of using a diary collection method compared with a CATI. Diary responses are not subject to recall bias (assuming prospective completion) hence it is probable that the diary information provides the more accurate figure, compared with the CATI. Recall bias may be responsible for either under or over-estimation as respondents may forget relevant episodes or they may report an episode from outside the period of interest as if it had happened within the period (forward telescoping) or vice versa (backward telescoping) [13]. In this study, the CATI gave the higher figure for toilet flushing (this figure was also influenced by conversion from an individual estimate to a household one) and lower figures for garden watering frequency and number of machine washing loads. In addition, the diary format has the advantage that it provides continuous numerical, rather than categorical, data for use in QMRA modelling. In contrast, the CATI for some water-using activities provided only categorical data.

Logistics and Cost

Given advantages conferred by use of a diary, rather than telephone interview for obtaining household exposure information, this leads to questions about how to overcome obstacles such as cost and logistics, commonly associated with diary administration. In this study, the estimated number of input researcher hours per completed diary was 2-3 times greater than that for each completed CATI (Table 1). The time estimate for the completion of each was up until statistical analysis of data was performed but was not inclusive of the initial recruitment process which was common to both the CATI and the diary; diary recruitment being dependent upon CATI uptake. The researcher time input per CATI was approximately 30 minutes, based on the interview time (15-20 min), look-up of household details and telephone contact information prior to the telephone call and verification of data (data was checked immediately after the interview).

The CATI time estimate assumes that one telephone call will result in a completed CATI. In fact this was not the case as some households required more than one call to be made before a CATI (44%) was generated (or refused) and a percentage of calls did not yield a CATI (61%)[15]. However, the assumption that one telephone call yields one CATI allows a more equitable comparison with the researcher time input for data preparation and handling following diary completion. This is because the starting point for the diary administration was agreement by the CATI respondent (70% agreed) that they wished to complete a water activity diary. Thus, each diary pack sent was expected to yield a completed diary that was returned to researchers. In fact this was not the case as the return rate (within the required 4 week time frame) of diaries was 63%. Based on this return rate and using a target of 200 completed diaries, the researcher time estimate per completed diary is approximately 85 minutes. This time estimate (expressed per unit diary) is based on: compilation of 320 diary packs, follow-up phone calls to 320 households to verify diary receipt; checking of completeness and legibility of 200 completed diary packs following receipt, tallying of data prior to entry onto data base, manual transfer of data to databases and data checking and verification.

A notable difference between the CATI and diary was that an incentive payment was offered for completion of the diary but not for completion of the CATI, adding to the budget costs for the diary. The use of a monetary incentive has been successful in other Australian studies [10] and was considered appropriate to compensate householders for recording their water usage over a 7-day recording period (a significant time commitment for the respondent) but not for a 15-20 minute CATI.

In this study, manual entry of data was performed, adding to the total researcher time input required for data entry to the diary database. However, it is possible that this time input could have been reduced if data scanning of diary records had been employed, reducing both the time requirement for data transfers into the ACCESS database and for data checking. Clearly a change such as this would require custom design of data input sheets and the corresponding data base. Whilst this would add to the cost of diary data base design, it is likely that it would not exceed that of the CATI database design. However, even if data scanning was implemented, the overall cost of diary administration would nonetheless exceed that of a CATI based on the cost of incentive payments and remaining labour costs associated with the preparation, sending and sorting of diary recording sheets.

A notable observation of this study was the quality of diary card completion, indicating that the diary cards were well designed (based on cards used in a prior Australian study [6]) and were easy to complete. The fact that there was no contact with households sent diary cards except for an initial phone call confirming diary card receipt and reinforcing that diary cards should be completed within a four week period was also testament to the clarity of diary instructions. This illustrates that it is possible to 'remotely' coordinate diary completion providing that adequate instructions are given and that diary recording forms are well designed. These observations are not only relevant to diary cards mailed to participants, but by extrapolation are also pertinent to web-based diary completion.

Limitations

This study was subject to a number of limitations including the sequential recruitment strategy in which householders were firstly recruited for the CATI and then (once the CATI was completed) were recruited to complete a 7-day water activity diary. This meant that the CATI 7-day period, always preceded the 7-day recording period for the diary. This limitation was countered as best as possible by confining the survey period to a maximum of 12 weeks (the summer period in which maximum household water consumption occurs). In addition, to reduce the time period between CATI and diary completion, a monetary incentive was offered to those households returning a completed diary within a 4 week period. Despite these measures, there was an approximate two to three week lag between CATI completion and diary commencement. The implications of this lag period are considered to be minimal for indoor water uses, which generally remain constant irrespective of weather conditions. However, this lag period potentially impacts on outdoor water uses such as garden watering where rainfall events and variations in temperature can influence garden irrigation frequency and duration. Rainfall in each of the CATI and diary 7-day periods was not tracked for individual households; hence the role of weather conditions on poor agreement between data collection tools for garden watering frequency was not able to be elucidated.

Another consequence of sequential recruitment was that only a subset of the household population that completed CATIs also completed diaries. It is therefore possible that the household population completing both the CATI and diary may have differed in some way from those that completed the CATI only. It is notable however that there was a high preparedness for householders to complete the diary (70% agreed) and that 63% of those agreeing to complete the diary also returned the diary within 4 weeks (a higher return rate may have resulted had the diary return been extended). The high rates of agreement to complete the diary support the view that financial reward was not a defining factor separating households that completed the diary from those that did not. Furthermore, the recycled water area in which householders were located is a niche housing development located in an area of high socio-economic status and is relatively homogenous with respect to household demographics and income.

Whilst it can be argued that it is likely that the diary population is representative of the larger household population completing the CATI only, a further question is whether the population completing the CATI are representative of the whole population of Rouse Hill dual reticulation households, including those that did not complete the CATI. In this study, CATI response rates were dependent upon the rate of matching of EWP and AEC records. The limiting factor in relation to household coverage was that only 57% of households were listed in the EWP, despite AEC records providing 87% of household coverage [15]. While it is possible that the water-using behaviour of listed and non-listed households in the EWP does vary, we consider that EWP listing is unlikely to be a primary determinant in the volume of water used by households. The relative uniformity in Rouse Hill dual household characteristics that drive water usage such as garden area, garden age and household size in the entire target survey area independent of EWP listing supports this assumption. Nonetheless, the possibility that the subset of households completing the diary and/or CATI were different from other households, based on a greater interest in water sustainability and advocates of higher recycled water use (and thus with a greater preparedness to devote time to diary and/or CATI completion), cannot be discounted.

Another limitation of this study was that there was no independent measure of the frequency and duration of individual water-using activities; hence the validity of both data collection tools is uncertain. It is therefore possible that both questioning via the CATI and recording of water usage using the diary were potentially subject to respondents reporting aspirational, rather than actual, behaviour. This is somewhat unlikely as this study was undertaken independently of the water authority supplying recycled and drinking water to householders and there was no disincentive for householders to report their actual water behaviour. Also, even though there were water restrictions relating to the use of drinking water outside the home during the study period they did not apply to inside use or to recycled water use; encompassing those water uses under investigation in this study.

Even though there were no adverse ramifications for the householder associated with reporting actual water usage, it is possible that drinking water restrictions may have influenced recycled water use despite lack of restrictions for recycled water. Such behavioural modifications are not predictable and are best addressed by the collection of contemporaneous water usage data. In doing this, the impact on exposure profiles, of behavioural changes associated with changing community attitudes and newly introduced water-using appliances, can be assessed.

Conclusion

The collection of water usage data is important to address data gaps for the assessment of the distribution of exposure of the human population in specific localities to contaminants in water through non-potable water use. This study showed that it is possible to successfully 'remotely' coordinate diary completion providing that adequate instructions are given and that diary recording forms are well designed. In addition, good diary return rates can be achieved using a monetary incentive and the diary format allows for collective recording, rather than an individual's estimation, of household water usage. Given reduced response rates associated with telephone databases, possible future alternative methods for data collection using diaries which by-passes telephone databases is to approach potential households by mail or via the Internet (as household Internet access increases). When exploring these options, particular attention should be paid to response rates in addition to diary costs, which in this study were greater than for the CATI.

Competing Interests

The authors declare that they have no competing interests.

Authors' Contributions

JO was responsible for the administration of both the diary and telephone surveys, design of the paper and analysis and interpretation of data. MS and KL were involved in the design concept of the project, drafting and revision of the final manuscript for intellectual content. All authors had full access to data in the study and had final responsibility for the decision to submit for publication. All authors have read and approved the final manuscript.

Acknowledgements

The authors would like to thank the Naomi Cooke for her technical assistance and statisticians at the Department of Epidemiology and Preventive Medicine, Monash University for their advice. We would also like to thank Sydney Water Corporation for their assistance during the conduct of the project and the Cooperative Research Centre for Water Quality and Treatment, Australia for funding the study.

References

1. NHMRC: Australian Drinking Water Guidelines. In National Water Quality Management Strategy. Canberra, ACT: National Health and Medical Research Council; 2004.

2. WHO: Guidelines for drinking-water quality. Volume 1 Recommendations. Geneva: World Health Organisation; 2004.

3. NRMMC/EPHC/AHMC: Australian Guidelines for Water Recycling: Managing Health and Environmental Risks (Phase 1). National Resource Management Ministerial Council; Environmental Protection and Heritage Council; and Australian Health Ministers Conference, Canberra; 2006.

4. Haas CN, Rose JB, Gerba CP, (Eds): Quantitative Microbial Risk Assessment. New York: John Wiley & Sons, Inc; 1999.

5. Streiner DL, Norman GR: Validity. In Health Measurement Scales: A Practical Guide to Their development and Use. 2nd edition. New York: Oxford University Press; 1995:145–162.

6. Metropolitan Water Authority: Domestic water use in Perth, Western Australia. Perth: Metropolitan Water Authority; 1985.

7. Fleiss JL: Statistical methods for rates and proportions. 2nd edition. New York: Wiley; 1981.

8. Bland JM, Altman DG: Statistical Methods for Assessing Agreement Between Two Methods of Clinical Measurement. Lancet 1986, 307–310.

9. Loh M, Coghlan P: Domestic water use study. Perth: Water Corporation, WA; 2003.

10. Roberts P: Yarra Valley Water 2003 Appliance Stock and Usage Patterns Survey. Melbourne: Yarra Valley Water; 2004.

11. Kaur S, Nieuwenhuijsen MJ, Ferrier H, Steer P: Exposure of pregnant women to tap water related activities. Occup Environ Med 2004, 61(5):454–460.

12. Barbone F, Valent F, Brussi V, Tomasella L, Triassi M, Di Lieto A, Scognamiglio G, Righi E, Fantuzzi g, Casolari L, et al.: Assessing the Exposure of Pregnant Women to Drinking Water Disinfection Byproducts. Epidemiology 2002, 13(5):540–544.

13. Tsui ELH, Leung G, Woo PPS, Choi S, Lo S-V: Under-reporting of inpatient services utilisation in household surveys – a population-based study in Hong Kong. BMC Health Serv Res 2005, 5(1):31.

14. Westrell T, Andersson Y, Stenstrom TA: Drinking water consumption patterns in Sweden. J Water Health 2006, 4(4):511–522.

15. O'Toole J, Sinclair M, Leder K: Maximising response rates in household telephone surveys. BMC Med Res Methodol 2008, 8:71.

Development of Scales to Assess Children's Perceptions of Friend and Parental Influences on Physical Activity

Russell Jago, Kenneth R. Fox, Angie S. Page, Rowan Brockman, and Janice L. Thompson

ABSTRACT

Background

Many children do not meet physical activity guidelines. Parents and friends are likely to influence children's physical activity but there is a shortage of measures that are able to capture these influences.

Methods

A new questionnaire with the following three scales was developed: 1) Parental influence on physical activity; 2) Motives for activity with friends scale;

and 3) Physical activity and sedentary group normative values. Content for each scale was informed by qualitative work. One hundred and seventy three, 10-11 year old children completed the new questionnaire twice, one week apart. Participants also wore an accelerometer for 5 days and mean minutes of moderate to vigorous physical activity, light physical activity and sedentary time per day were obtained. Test-retest reliability of the items was calculated and Principal Component analysis of the scales performed and sub-scales produced. Alphas were calculated for main scales and sub-scales. Correlations were calculated among sub-scales. Correlations between each sub-scale and accelerometer physical activity variables were calculated for all participants and stratified by sex.

Results

The Parental influence scale yielded four factors which accounted for 67.5% of the variance in the items and had good ($\alpha > 0.7$) internal consistency. The Motives for physical activity scale yielded four factors that accounted for 66.1% and had good internal consistency. The Physical activity norms scale yielded 4 factors that accounted for 67.4% of the variance, with good internal consistency for the sub-scales and alpha of .642 for the overall scale. Associations between the sub-scales and physical activity differed by sex. Although only 6 of the 11 sub-scales were significantly correlated with physical activity there were a number of associations that were positively correlated >0.15 indicating that these factors may contribute to the explanation of children's physical activity.

Conclusion

Three scales that assess how parents, friends and group normative values may be associated with children's physical activity have been shown to be reliable and internally consistent. Examination of the extent to which these new scales improve our understanding of children's physical activity in datasets with a range of participant and family characteristics is needed.

Background

Regular physical activity has many short and long-term benefits for children including lower body mass index [1] and lower mean values for cardiovascular risk factors [2-4]. Physical activity is also associated with higher levels of mental well-being among children [5] and helps children to develop social skills [6,7]. Despite these benefits many children and adolescents do not engage in recommended amounts of physical activity [8,9]. The mediating variable model suggests that in order to change a behavior (such as physical activity) we need to change the key factors or mediators of that behavior [10]. Therefore, in order to develop effective interventions to increase children's physical activity we need to understand the factors, or correlates of children's physical activity and then change those variables [11].

Parents and friendship groups are likely to be key influences on children's physical activity behaviors. Davison and colleagues reported that fathers' modeling of active behaviors and mothers' logistic support for physical activity (e.g., enrolling child in sport programs and going to sporting events with the child) were associated with the physical activity levels of 9 year old girls [12]. Salvy and colleagues reported that 10 year old boys and girls were more likely to engage in high intensity physical activity with friends [13] and that friends increased children's motivation for physical activity. Peer support for physical activity has also indicated an association with higher amounts of physical activity among fifth to eighth grade US students [14]. While these studies demonstrate the importance of the influence of parents and friends on children's physical activity they provide limited information about how these influences are manifested. More information about potential mechanisms through which interactions with parents and friends shape physical activity behaviors is therefore needed to develop effective strategies to increase children's physical activity.

Our research team have conducted extensive qualitative work to examine how friends and parents influence the physical activity behaviors of 10-11 year old children [15-17]. The data on the influence of friends showed that 10-11 year old British children have three types of friendship groups: school friends, neighborhood friends and other friends (e.g. children of their parents' friends and children from youth or community groups) with most children belonging to multiple groups [15]. Findings also suggested that friendship group members shared common attitudes and perceived normative values for physical activity and screen-viewing [15]. Children also reported that their reasons or motives for participating in physical activity included intrinsic appeal, increasing social affiliation and preventing isolation [15]. Motives for physical activity participation were also influenced by group affiliation [15]. Therefore, understanding group affiliation may be important for understanding physical activity attitudes, expectations and norms among children.

Our qualitative work indicated that parents exert considerable influence on children's behavior through provision of transport assistance, financial support, modeling, encouragement, and setting rules for activity [16,17]. Several participants reported that their family structure and particularly the parents or guardians who they live with differ for weekday and weekend days. This suggests that measures should accommodate these differences when examining parental influence.

There is a need to understand how friends and parental factors influence children's physical activity, parental rules for physical activity, and physical activity and screen-viewing subjective normative values. Subjective normative values are a key component of the Theory of Planned Behavior and represent a person's perception of other peoples (i.e. other children's) preferences for engaging in a behavior such as physical activity [18]. However, although there are reliable measures that capture parental support for physical activity [12] there are no scales that also examine parent imposed activity-related rules, or scales that can address how associations may differ by weekday or weekend parent. Similarly, while some measures have included items that address the extent to which spending time with friends contribute to physical activity enjoyment [19] to our knowledge no current measure identifies friend influences on physical activity and particularly if influence differs by type of friend. Finally, although physical activity norms and motives for physical activity scales have been developed these scales have tended to focus on more global subjective norms and included questions on fellow students, teacher and parent norms [20] and have not included screen-viewing behaviors. To address these limitations this paper reports the development and reliability assessment of three new questionnaire scales to assess parental influence on physical activity, motives for activity with friends, and group norms related to physical activity and sedentary behaviors.

Methods

Participants were 173, 10-11 year old children recruited from 7 primary schools in Bristol, England. We initially approached 9 primary schools with 2 schools declining, one because of recent changes to the school management team and the other due to participation in a number of other, non-physical activity based research projects. There were 373, Year 6 pupils within the 7 primary schools and as such the recruitment rate was 46.4%. Participant sex and highest education within the household were obtained by parental report. The study was approved by a University of Bristol ethics committee and informed consent and assent were obtained for all participants [21].

Questionnaire Development

The questionnaire included three main scales: 1) Parental influence on physical activity scale which included questions about the parents that children spent time with on both weekdays and weekend days; 2) Motives for activity with friends scale which examined reasons for participating in physical activity and whether motives were different for school friends, neighborhood friends or other friends;

and 3) Physical activity and sedentary group normative (norms) values. Content for each scale was derived from our qualitative work with these populations [15-17]. For example, a number of the children who took part in focus groups reported that social factors, prevention of bullying and isolation and spending time with friends influenced their participation in physical activity with participants also expressing diverse views about the merits and perceived negative connotations of participation in team sports, and screen-viewing behaviors. Items that were designed to capture all of these issues were generated by the first author and then reviewed, modified and added to by all of the co-authors. This process was repeated several times until all of the study team were satisfied with the items. Items were phrased as statements to which participants were provided with four response options: disagree a lot; disagree; agree; and agree a lot. The response options were modeled after the responses for an Australian survey [22] which included two additional options for neither (disagree or agree) or don't know. These two options were omitted from the survey as our pilot work indicated that UK children found these two options confusing.

Procedures

Height was measured to the nearest mm with a SECA Leicester Stadiometer. Weight was measured to the nearest 0.1 kg using a SECA 899 digital scale. Body mass index (BMI) was calculated (kg/m^2). Participants completed the new questionnaire twice, with the second administration approximately a week after the first. Physical activity was objectively measured using an accelerometer (Actigraph GT1M; Actigraph LLC USA) programmed to record data every 10 seconds. Children wore the accelerometer on a belt around their waist during waking hours for five consecutive days, including weekends. At the end of the measurement period the monitors were collected by researchers and the data downloaded to a PC. Any 20 minute periods of zero activity recorded by the accelerometer were taken to indicate that the accelerometer was not being worn and were classified as missing data. Participants were included in the analysis if they provided at least 500 minutes of data for at least 3 days. It has previously been reported that between 3 and 7 days of accelerometer data are required to provide an indication of habitual physical activity using accelerometers [23-25]. Therefore, we employed a three day inclusion criteria as the minimum threshold for our data. While we accept that it has been suggested that more than 3 days might be needed to capture the less predictable behavior of children [23,25], the 3-day inclusion criteria has been widely used for children and adolescents [26-28] and provided us with the largest possible sample size. Mean counts per minute were calculated to provide an indication of the overall volume of physical activity in which the participants engaged. Mean minutes of sedentary, light and moderate to vigorous physical activity per day were also calculated using child-specific cut-points [29]. However, as the count values derived from the GT1M are 9% higher than those obtained from the original 7164 monitors which were used to derive intensity thresholds, a correction factor of 0.91 was used for all intensity values [30].

Statistical Analyses

Descriptive statistics including means, standard deviations and percents were calculated for all demographic and physical activity variables. Initial checks indicated that there was variance in responses to each item. As the aim of this paper was to develop questionnaires with good test-retest reliability, paired t-tests were used to assess the test-retest reliability of individual items prior to factor analysis. Items that were significantly different (P <.001) between the two administrations were omitted from further analysis. Pearson correlations were then conducted between the first and second administrations of the items and any items that were not associated were not included in further analysis.

For items that were retained after the initial reliability analyses we also calculated test retest intra-class correlations (ICC). Although intra-class correlations have been frequently recommended for reliability studies [31-33] authors have commented on the number of different ICC's that could be used for reliability studies and the lack of a clear consensus on when and why to select a particular type of ICC [31,33]. We performed two-way random effect (subjects and time are both random) intra-class correlations that assessed absolute agreement (as ideally you would want the same response two weeks apart). There is also a debate about the criteria that should be used to assess the reliability ICC's. A number of authors [34,35] have applied the "benchmark" criteria of Landis and Koch that was initially described for Kappa statistics [36]. According to these criteria test re-test ICC's are interpreted as: 0.21 - 0.40 Fair agreement, 0.41 - 0.60 Moderate agreement, 0.61 - 0.80 Substantial agreement and 0.81 - 1.00 Almost perfect agreement [36]. In light of the uncertainty over how to apply these criteria we opted to remove any item that had an ICC that was less than 0.4. However, in light of the ambiguity on how to apply and interpret the ICC's we report the ICC's for each item retained in the analysis but did not apply any further exclusion criteria. In terms of interpretation ICC's of 0.4 -0.59 were considered acceptable but improvable, 0.60- 0.79 satisfactory, ≥ 0.80 excellent [35].

Once reliable items had been identified Principal Component analysis with Varimax rotation was then conducted separately for each of the three scales. The resulting scree plots and eigen-values were inspected, interpretability considered, and factors selected. Items that did not load on factors (at least 0.4), or loaded on multiple factors were removed and the models re-run. Values for items included in rotated factors were summed and used in analyses. The internal consistency of all of the items included in each resulting factor was then assessed using Cronbach's Alpha. Alpha was then also calculated for all of the items that were retained in each of the three overall scales.

Bivariate Pearson correlations were used to examine inter-relationships between each of the factor scores that were derived from the Principal Component analysis. To provide an indication of the relevance of these measures in relation to physical activity bivariate Pearson correlations were then conducted between the factor scores and all four of the accelerometer derived physical activity variables. However, since extensive research has shown that children's physical activity [8] differ by sex and that the associations between psychosocial variables and children's physical activity differ by sex [37,38], correlations with physical activity were calculated for all participants and then stratified by sex.

Results

Demographic characteristics for the 173 participants in the study are shown in Table 1. The sample was 51% female with 47% living in households in which the highest level of education was GCSE (national school examinations assessed at age 16). Valid accelerometer data were obtained for 131 participants with the participants obtaining an average of 21.5 minutes of moderate to vigorous intensity physical activity per day.

Table 1. Participant characteristics

	n	%
Male	86	49.1
Female	89	50.9
Highest level of Education		
GCSE or similar	83	47.4
A'Level or similar	33	18.9
Degree	36	21.7
Higher degree	13	7.4
Missing	8	4.6
	n	Mean (SD)
BMI	173	19.0 (6.6)
Accelerometer minutes of Moderate to Vigorous Physical Activity per day	131	21.5 (12.3)
Accelerometer minutes of Light Activity per day	131	121.2 (25.6)
Accelerometer minutes of sedentary time per day	131	1217.3 (56.8)
Accelerometer counts per minute	131	457.7 (111.6)

The final Parental influence on physical activity scale is presented in Table 2. No items were dropped from the scale which included 14 items and accounted for 67.5% of the overall variance and had a reasonable internal consistency (alpha =.75).

Table 2. Parental influence on children's physical activity scale and factor structure

	General parenting support	Active parents	Past activity	Guiding support
PERCENT VARIANCE EXPLAINED	23.2%	19.4%	12.5%	12.4%
ALPHA FOR SUB-SCALE	.830	.836	.802	.819
The adult(s) I live with on a weekend day pay for me to take part in physical activity (for example paying for swimming or to attend football club)	.833	.041	.069	.038
The adult(s) I live with on a weekend day drive me to sports clubs	.774	-.056	.071	.133
The adult(s) I live with on a weekday pay for me to take part in physical activity (for example paying for swimming or to attend football club)	.760	.047	.027	-.115
The adult(s) I live with on a weekday take me to or collect me from sport or exercise clubs	.759	.085	-.177	.059
The adult(s) I live with on a weekend day encourage (or tell) me to be physically active	.632	.204	.017	-.006
The adult(s) I live with on a weekday encourage (or tell) me to be physically active	.577	.284	.074	-.044
The adult(s) I live with on a weekend day take part in lots of physical activity	.172	.812	-.216	.024
The adults(s) I live with on a weekend day take part in physical activity with me	.044	.810	.082	.002
The adult(s) I live with on a weekday take part in physical activity with me	.072	.787	.086	.119
The adult(s) I live with on a weekday take part in lots of physical activity	.145	.786	-.090	.037
The adult(s) I live with on a weekday used to take part in lots of physical activity but they don't anymore	.005	-.015	.906	.031
The adult(s) I live with on a weekend day used to take part in lots of physical activity but they don't anymore	.069	-.056	.888	.209
The adult(s) I live with on a weekday have rules for physical activity (such as being home at a set time, not going to some places etc)	-.007	.036	.099	.908
The adult(s) I live with on a weekend day have rules for physical activity (such as being home at a set time, not going to some places etc)	.038	.113	.120	.904

Alpha for overall Scale = .746   Overall variance explained = 67.5%

The General parenting support scale provides an indication of the overall support that the child perceives their parent provides for physical activity (23.2% of the variance, alpha =.83) while the Active parents sub-scale provides a measure of the extent to which the child perceives his or her parent to be active (19.4% of the variance, alpha =.84). Past parental activity provides an indication whether or not the child perceives that the parental used to be active (12.5% of the variance, alpha =.80) while the Guiding support (12.5% of the variance, alpha =.82) scale captures the extent to which the child's parents have supportive rules for physical activity participation. Intra-class correlations for all items included in the Parental influence on physical activity were between 0.6 and 0.8 (7 > 0.7) suggesting satisfactory reliability.

The final Motives for physical activity with friends scale is presented in Table 3. Three items were not included in the factor analysis due to poor reliability, with an additional four items removed from the scale because they cross-loaded onto multiple factors. The scale had 14 items that accounted for 66.1% of the overall variance and had good internal consistency (alpha = 0.86). The Prevent bullying sub-scale identifies the extent to which the child is motivated to engage in activity to prevent peer victimization (18.6% of the overall variance, alpha =.79). The Social sedentary sub-scale identifies the extent to which a child engages in sedentary behaviours for social reasons (16.8% of the variance, alpha =.76) while the Social affiliation sub-scale (15.6% of the variance alpha =.71) identifies the extent to which group affiliation influences activity participation. Finally, the Neighborhood friends sub-scale (15.2% of the variance, alpha =.79) identifies the extent to which children are specifically motivated to engage in physical activity to spend time with children who live in their neighborhood. The intra-class correlation for the first item on the Prevent Bullying sub-scale was excellent (.845), 10 further items had satisfactory ICC's (7 > 0.7) while the remaining 3 items had ICC's that were considered acceptable but improvable ['I take part in sitting down activity to spend time with my friends' (.522); 'I take part in physical activity because my friends at school do' (.504); and 'I take part in physical activity because my neighborhood friends do' (.564)].

Table 3. Motives for activity with friends' scale and factor structure

	Prevent bullying	Social Sedentary	Social affiliation	Neighborhood friends
PERCENT VARIANCE EXPLAINED	18.6%	16.8%	15.6%	15.2%
ALPHA FOR SUB-SCALE	.789	.761	.714	.793
I take part in physical activity with my other friends so that I don't get picked on	.822	.075	.254	.056
I take part in physical activity with my neighborhood friends so that I don't get picked on	.783	.117	.018	.264
I take part in physical activity at school so that I don't get picked on	.745	.112	.195	.171
I take part in sitting down activity because my other friends do	.510	.391	.216	.051
I take part in sitting down activity to spend time with my other friends	.108	.844	.102	.093
I take part in sitting down activity to spend time with my school friends	.009	.814	.138	.061
I take part in sitting down activity because I want to belong in a group with my other friends	.318	.684	.153	.213
I take part in physical activity because my friends at school do	.183	.076	.762	-.046
I take part in physical activity to spend time with my friends	-.027	.204	.759	.155
I take part in physical activity because my other friends do	.363	.017	.653	.099
I take part in physical activity because I want to belong in a group with my other friends	.284	.335	.519	.200
I take part in physical activity to spend time with my neighborhood friends	.077	.061	.132	.917
I take part in sitting down activity to spend time with my neighborhood friends	.178	.395	-.118	.748
I take part in physical activity because my neighborhood friends do	.321	.029	.302	.703

Alpha for overall Scale = .857   Overall variance explained = 66.1%

One item was removed from the final Physical activity and sedentary norms scale because the test retest ICC was low (.334) but all other original items were retained in a scale that accounted for 67.9% of the overall variance and had an alpha of .64 (Table 4). The Sedentary sub-scale identifies norms for screen-viewing behaviors (28.9% of the variance) with an Activity sub-scale providing comparable information for physical activity norms (17.6% of the variance). The analysis also yielded a Teasing sub-scale (20.8% of the variance) which provides information on the extent to which respondents feel that they would be teased for engaging in physical activity and screen-viewing. The ICC's for the retained variables indicated that 8 items had satisfactory reliability (>0.6) and 2 items [Kids my age spend lots of time watching TV or DVD's =.44, Kids my age think taking part in physical activity is a good thing to do =.55] had acceptable but improvable reliability.

Table 4. Physical activity norms scale and factor structure

	Sedentary	Teasing	Activity
PERCENT VARIANCE EXPLAINED	28.9%	20.8%	17.6%
ALPHA FOR SUB-SCALE	.795	.709	.701
Kids my age think watching TV/DVDs is a good thing to do	.829	-.014	-.260
Kids my age think playing computer games (such as XBOX, PlayStation or Nintendo) is a good thing to do	.830	-.062	-.170
Kids my age spend lots of time playing on games consoles (such as XBOX, PlayStation or Nintendo)	.765	.133	.275
Kids my age spend lots of time watching TV or DVD's	.710	.238	.213
Kids my age would tease me if I spent a lot of time taking part in physical activity	.179	.870	-.107
Kids my age would tease me if I went to lots of after-school sport or other sports clubs	.129	.859	-.142
Kids my age would tease me if I spend a lot of time playing computer games (such as XBOX, PlayStation or Nintendo)	-.075	.734	.066
Kids my age think attending after-school or sports clubs is a good thing	.012	-.124	.783
Kids my age think taking part in physical activity is a good thing to do	-.132	-.178	.807
Kids my age take part in lots of physical activity	.104	.123	.701
Alpha for overall Scale = .642		**Overall variance = 67.4%**	

The associations among the 11 sub-scales are presented in Table S1 (Additional File 1). There were a number of statistically significant associations among the sub-scales with the strongest associations within the same scale being the Avoid bullying and Social affiliation sub-scales of the motives for activity with friends scale (r =.521, p <.001). There were also significant associations between sub-scales derived from different main scales for example Teasing norms was significantly associated with the Avoid bullying sub-scale of the motives for activity overall scale (r =.427, p <.001).

Correlation analyses indicated that when the sample included all participants, the Active parent sub-scale of the parental influence scale was associated with light intensity physical activity (r =.178, p =.042) while Active norms was associated with minutes of MVPA per day (r =.181, p =.039) (data not in tabular form). Pearson correlations between each of the 11 sub-scales and the four physical activity variables are presented separately for boys and girls in Table S2 (Additional File 2). Among girls, Social sedentary was associated with sedentary minutes per day (r =.273, p =.024) and negatively associated with mean counts per minute (r = -.249, p =.040). Parental past activity was negatively associated with minutes of light activity per day (r = -.307, p =.011) and Avoid bullying was negatively associated with minutes of MVPA per day (r = -.255, p =.037) and accelerometer counts per minute (r = -.268, p =.028). For boys, Social affiliation was associated with minutes of MVPA per day (r =.253, p =.050).

Discussion

In this paper we have presented information on the factor structure and reliability of three new scales: the Parental influence on physical activity, Motives for activity with friends and Physical activity and sedentary norms. Items were only included in the scales if they had acceptable test - retest reliability and variance in responses and thus all scales can be considered reliable. The alphas for the Parental influence on physical activity and Motives for activity with friends scales as well as the alphas for all of the sub-scales of these measures were >0.7. Alpha values >0.7 are considered satisfactory for non-clinical instruments [39] and therefore we can be confident that the items included in these scales were measuring coherent concepts. The alpha for the overall Norms scale was.64 indicating that caution is required when attempting to use all of the items on this overall scale to describe friend related physical activity norms. Collectively, these analyses therefore highlight that we have developed new scales that have good test re-test reliability and internal consistency.

The associations between the sub-scales and physical activity were different for boys and girls, suggesting that the extent to which new sub-scales predicted physical activity differed by sex. For example, the correlation between Parental past activity and light intensity physical activity was -.307 for girls but there was no association for boys. This would suggest that parental physical activity influences girl's physical activity only. Similarly, the Social sedentary scale correlated.275 with minutes of sedentary time for girls but there was no association for boys. These findings are consistent with previous research which has shown that the association between correlates of children's physical activity differs by gender. For example, gender differences in the associations between self-efficacy, social norms, beliefs, and outcomes of children and adolescents have been reported [37,40]. Findings suggest that the parental and friendship factors derived from our new questionnaire could be important predictors of behavior, but associations may well be sex specific and thus further research that examines these differences is required.

Although not statistically significant (p <.05), a number of the sex stratified associations between the sub-scales and physical activity variables were in excess of 0.15 and often above 0.20. Such associations are comparable to the associations between physical activity self-efficacy and physical activity [40-42]. A number of physical activity interventions have been designed in which self-efficacy is hypothesized to be a key mediator of physical activity behavior change [43,44]. Although there is a shortage of studies that have employed mediating variable analyses of self-efficacy based interventions [11,45] this is likely to be a function of the lack of success in changing youth physical activity which hampers the detection of mediation effects [45]. The comparison is salient because it suggests that these sub-scales could explain a considerable amount of the variance in children's physical activity behaviors. Therefore an appropriately powered study is needed to fully examine the associations between these constructs and children's physical activity.

The Prevent bullying sub-scale accounted for the highest amount of variance (18.6%) of the four factors on Motives for activity with friends scale, suggesting that this concept is particularly salient for some participants. Interestingly, Teasing was a specific sub-scale on the Norms scale suggesting that this factor was also a salient normative value. Being physically active is associated with a reduction in the likelihood that 13-15 year old adolescents are bullied [46]. Furthermore, athletic identity is associated with increased physical activity among children [47] and many children, particularly girls, are socialized out of physical activity by teasing or peer victimization on the basis of poor sporting ability [48]. As such these two sub-scales may identify children who are concerned about or perhaps at risk of peer teasing. The scales could also be utilized as a means of identifying particular groups of children who do not engage in key behaviors because of fears of social isolation and teasing.

The General parenting support and Active parents sub-scales include similar items to Davison's logistic support factor [12] but utilize more specific examples and have resulted in two factors rather than one. An interesting area of future research would therefore be to consider how the new sub-scales and Davison's measure are related to each other and whether utilizing all scales increases our understanding of how parents can help to support physical activity.

The Neighborhood friends sub-scale suggests that there is something specific about how children identify with this group of friends. As the concept of neighborhood friends and their influence on physical activity behaviors is new [15], exploring the role of this friendship group and how neighborhood friends can help promote activity in less active children is likely to be essential in fully understanding why different groups of children are active.

In the Motives for Activity with Friends scale there were two social sub-scales: Social sedentary which captured preferences for engaging in sedentary behaviors with friends; and Social affiliation which captured engaging in activity to spend time with friends. A number of studies have reported that social factors are associated with participation in physical activity [15,49-51], and these scales extend that work by indicating that the social aspects of screen-viewing and physical activity are likely to be different. The increased specificity of these sub-scales suggest that they may be able to explain more of the variance in the behaviors to which they relate than current measures and research that focuses on this possibility is needed. Moreover, like the teasing and prevent bullying sub-scales, these two new sub-scales may be useful in developing profiles of children, particularly children who have preferences for either physical activity or screen-viewing behaviors.

The analysis presented in this paper has focused on the reliability of the new scales and not the "validity" of the scales [52]. However, as the development of the scales and items were informed by extensive qualitative work and the items assess the issues raised in the qualitative work the items can be considered to have "face and content validity" [52]. The items included in the scales address new constructs that were identified in the qualitative work and which have not been reported before and therefore there is no existing scale against which to compare these items. As such we are unable to assess the "criterion validity" but in order to provide an indication of the potential utility of these scales we have provided information on the associations with physical activity the key behavior to which they are hypothesized to relate.

Strengths/Limitations

This study has developed and provided reliability information on new scales that provide information on how friends and peers influence children's physical activity patterns. However, while we have been able to demonstrate the reliability of these measures, the relatively small sample limits our ability to examine associations with physical activity and particularly limits our ability to examine sex-specific associations. Moreover, as we did not collect family structure data we are unable to examine if responses for the new Parental Influence scale which assesses the influence of weekday and weekend parents differed by the time that children spent with different parents. It is also important to recognize that the development and piloting of this questionnaire has been conducted in one area of the United Kingdom, and as such the concepts assessed could be influenced by the location in which the children reside and may require refinement for use in other countries. Moreover, the development and factor analysis has only been conducted in a single sample and, as such, confirmation of the factor structure in another sample may be required before widespread adoption.

Conclusion

Three scales that assess how parents, friends and group normative values may be associated with children's physical activity have been shown to be reliable and internally consistent. Initial analyses suggests that these measures will provide new information on the factors that influence children's physical activity but more research in a larger dataset is required to identify how associations may differ by sex and participant characteristics. They also provide further indication of the importance of conceptualizing physical activity as social-context specific with different parental and peer influences in action in different settings and time periods in the week.

Competing Interests

The authors declare that they have no competing interests.

Authors' Contributions

This paper was conceived and drafted by RJ. All authors provided critical input into the design of the research, drafting of the paper and provided edits to the paper. All authors read and approved the final manuscript.

Acknowledgements

This project was funded by a project grant from the British Heart Foundation (ref PG/06/142). We would like to thank all of the children and schools who participated in this study.

References

1. Jago R, Baranowski T, Baranowski JC, Thompson D, Greaves KA: BMI from 3-6 y of age is predicted by TV viewing and physical activity, not diet. Int J Obes Relat Metab Disord 2005, 29:557–564.

2. Jago R, Wedderkopp N, Kristensen PL, Moller NC, Andersen LB, Cooper AR, Froberg K: Six-year change in youth physical activity and effect on fasting insulin and HOMA-IR. Am J Prev Med 2008, 35:554–560.

3. Brage S, Wedderkopp N, Ekelund U, Franks PW, Wareham NJ, Andersen LB, Froberg K: Objectively measured physical activity correlates with indices of insulin resistance in Danish children. The European Youth Heart Study (EYHS). Int J Obes Relat Metab Disord 2004, 28:1503–1508.

4. Brage S, Wedderkopp N, Ekelund U, Franks PW, Wareham NJ, Anderson LB, Froberg K: Features of the metabolic syndrome are associated with objectively measured physical activity and fitness in Danish children: the European Youth Heart Study. Diabetes Care 2004, 27:2141–2148.

5. Parfitt G, Eston RG: The relationship between children's habitual activity level and psychological well-being. Acta Paediatr 2005, 94:1791–1797.

6. Bailey R: Evaluating the relationship between physical education, sport and social inclusion. Educational Review 2005, 57:71–90.

7. Hansen DM, Larson RW, Dworkin JB: What adolescents learn in organized youth activities: A survey of self-reported developmental experiences. Journal of Research on Adolescence 2003, 13:25–55.

8. Jago R, Anderson C, Baranowski T, Watson K: Adolescent patterns of physical activity: Differences by gender, day and time of day. Am J Prev Med 2005, 28:447–452.

9. Riddoch CJ, Mattocks C, Deere K, Saunders J, Kirkby J, Tilling K, Leary SD, Blair SN, Ness AR: Objective measurement of levels and patterns of physical activity. Arch Dis Child 2007, 92:963–969.

10. Baranowski T, Anderson C, Carmack C: Mediating variable framework in physical activity interventions. How are we doing? How might we do better? Am J Prev Med 1998, 15:266–297.

11. Baranowski T, Jago R: Understanding mechanisms of change in children's physical activity programs. Exercise and Sport Science Reviews 2005, 33:163–168.

12. Davison KK, Cutting TM, Birch LL: Parents' activity-related parenting practices predict girls' physical activity. Med Sci Sports Exerc 2003, 35:1589–1595.

13. Salvy SJ, Bowker JW, Roemmich JN, Romero N, Kieffer E, Paluch R, Epstein LH: Peer influence on children's physical activity: an experience sampling study. J Pediatr Psychol 2008, 33:39–49.

14. Beets MW, Vogel R, Forlaw L, Pitetti KH, Cardinal BJ: Social support and youth physical activity: the role of provider and type. Am J Health Behav 2006, 30:278–289.

15. Jago R, Brockman R, Fox KR, Cartwright K, Page AS, Thompson JL: Friendship groups and physical activity: qualitative findings on how physical activity is initiated and maintained among 10-11 year old children. Int J Behav Nutr Phys Act 2009, 6:4.

16. Jago R, Thompson JL, Page AS, Brockman R, Cartwright K, Fox KR: Licence to be active: parental concerns and 10-11-year-old children's ability to be independently physically active. J Public Health (Oxf) 2009, in press.

17. Brockman R, Jago R, Fox KR, Thompson JL, Cartwright K, Page AS: Get off the sofa and go and play": family and socioeconomic influences on the physical activity of 10-11 year old children. BMC Public Health 2009, 21:253.

18. Conner M, Sparks P: Theory of Planned Behaviour and Health Behaviour. In Predicting health behaviour. Edited by: Conner M, Norman P. London: Open University Press; 2005:170–222.

19. Motl RW, Dishman RK, Saunders R, Dowda M, Felton G, Pate RR: Measuring enjoyment of physical activity in adolescent girls. Am J Prev Med 2001, 21:110–117.

20. Motl RW, Dishman RK, Trost SG, Saunders RP, Dowda M, Felton G, Ward DS, Pate RR: Factorial validity and invariance of questionnaires measuring social-cognitive determinants of physical activity among adolescent girls. Prev Med 2000, 31:584–594.

21. Jago R, Bailey R: Ethics and paediatric exercise science: Issues and making a submission to a local ethics and research committee. Journal of Sport Sciences 2001, 19:527–535.

22. Timperio A, Ball K, Salmon J, Roberts R, Giles-Corti B, Simmons D, Baur LA, Crawford D: Personal, family, social, and environmental correlates of active commuting to school. Am J Prev Med 2006, 30:45–51.

23. Baranowski T, Masse LC, Ragan B, Welk G: How many days was that? We're still not sure, but we're asking the question better! Med Sci Sports Exerc 2008, 40:S544–549.

24. Matthews CE, Ainsworth BE, Thompson RW, Bassett D: Sources of variance in daily physical activity levels as measured by an accelerometer. Med Sci Sports Exerc 2002, 34:1376–1381.

25. Trost SG, Pate RR, Freedson PS, Sallis JF, Taylor WC: Using objective physical activity measures with youth: How many days are needed. Med Sci Sports Exerc 2000, 32:426–431.

26. Cooper AR, Wedderkopp N, Jago R, Kristensen PL, Moller NC, Froberg K, Page AS, Andersen LB: Longitudinal associations of cycling to school with adolescent fitness. Prev Med 2008, 47:324–328.

27. Ekelund U, Anderssen S, Andersen LB, Riddoch CJ, Sardinha LB, Luan J, Froberg K, Brage S: Prevalence and correlates of the metabolic syndrome in a population-based sample of European youth. Am J Clin Nutr 2009, 89:90–96.

28. Sardinha LB, Baptista F, Ekelund U: Objectively measured physical activity and bone strength in 9-year-old boys and girls. Pediatrics 2008, 122:e728–736.

29. Puyau MR, Adolph AL, Vohra FA, Butte NF: Validation and calibration of physical activity monitors in children. Obes Res 2002, 10:150–157.

30. Corder K, Brage S, Ramachandran A, Snehalatha C, Wareham N, Ekelund U: Comparison of two Actigraph models for assessing free-living physical activity in Indian adolescents. J Sports Sci 2007, 25:1607–1611.

31. Atkinson G, Nevill AM: Statistical methods for assessing measurement error (reliability) in variables relevant to Sports Medicine. Sports Med 1998, 26:217–238.

32. Yen M, Lo LH: Examining test-retest reliability: an intra-class correlation approach. Nurs Res 2002, 51:59–62.

33. Batterham AM, George KP: Reliability in evidence-based clinical practice: a primer for allied health professionals. Physical Therapy in Sport 2003, 4:122–128.

34. Brown WJ, Trost SG, Bauman A, Mummery K, Owen N: Test-retest reliability of four physical activity measures used in population surveys. J Sci Med Sport 2004, 7:205–215.

35. March JS, Sullivan K: Test-retest relaibility of the mutidimensional anxiety scale for children. J Anxiety Disord 1999, 13:349–358.

36. Landis JR, Koch GG: The measurement of obsever agreement for categorical data. Biometrics 1977, 33:159–174.

37. Trost SG, Pate R, Ward DS, Saunders R, Riner W: Correlates of objectively measured physical activity in preadolescent youth. Am J Prev Med 1999, 17:120–126.

38. Horst K, Paw MJ, Twisk JW, Van Mechelen W: A brief review on correlates of physical activity and sedentariness in youth. Med Sci Sports Exerc 2007, 39:1241–1250.

39. Bland JM, Altman DG: Cronbach's alpha. Bmj 1997, 314:572.

40. Jago R, Baranowski T, Watson K, Bachman C, Baranowski JC, Thompson D, Hernandez AE, Venditti E, Blackshear T, Moe E: Development of new physical activity and sedentary behavior change self-efficacy questionnaires using item response modeling. Int J Behav Nutr Phys Act 2009, 6:20.

41. Saunders RP, Pate RR, Felton G, Dowda M, Weinrich MC, Ward DS, Parsons MA, Baranowski T: Development of questionnaires to measure psychosocial influences on children's physical activity. Prev Med 1997, 26:241–247.

42. Ryan GJ, Dzewaltowski DA: Comparing the relationship between different types of self-efficacy and physical activity in youth. Health Educ Behav 2002, 29:491–504.

43. Jago R, Baranowski T, Baranowski J, Thompson D, Cullen K, Watson K, Liu Y: Fit for life Boy Scout badge: Outcome Evaluation of a troop & internet intervention. Prev Med 2006, 42:181–187.

44. van Sluijs EM, McMinn AM, Griffin SJ: Effectiveness of interventions to promote physical activity in children and adolescents: systematic review of controlled trials. Bmj 2007, 335:703.

45. Baranowski T, Cerin E, Baranowski J: Steps in the design, development and formative evaluation of obesity prevention-related behavior change trials. Int J Behav Nutr Phys Act 2009, 6:6.

46. Turagabeci AR, Nakamura K, Takano T: Healthy lifestyle behaviour decreasing risks of being bullied, violence and injury. PLoS ONE 2008, 3:e1585.

47. Anderson CB, Coleman KJ: Adaptation and validation of the athletic identity questionnaire-adolescent for use with children. J Phys Act Health 2008, 5:539–558.

48. Kunesh MA, Hasbrook CA, Lewthwaite R: Physical Activity Socialization: Peer Interactions and Affective Responses Among a Sample of Sixth Grade Girls. Sociology of Sport Journal 1992, 9:385–396.

49. Price SM, McDivitt J, Weber D, Wolff LS, Massett HA, Fulton JE: Correlates of weight-bearing physical activity among adolescent girls: results from a national survey of girls and their parents. J Phys Act Health 2008, 5:132–145.

50. DiLorenzo TM, Stuckey-Ropp RC, Wal JS, Gotham HJ: Determinants of exercise among children. II. A longitudinal analysis. Prev Med 1998, 27:470–477.

51. Voorhees CC, Murray D, Welk G, Birnbaum A, Ribisl KM, Johnson CC, Pfeiffer KA, Saksvig B, Jobe JB: The role of peer social network factors and physical activity in adolescent girls. Am J Health Behav 2005, 29:183–190.

52. Coolican H: Research methods and Statistics in Psychology. second edition. London: Hodder & Stoughton; 1994.

Preliminary Spatiotemporal Analysis of the Association Between Socio-Environmental Factors and Suicide

Xin Qi, Shilu Tong and Wenbiao Hu

ABSTRACT

Background

The seasonality of suicide has long been recognised. However, little is known about the relative importance of socio-environmental factors in the occurrence of suicide in different geographical areas. This study examined the association of climate, socioeconomic and demographic factors with suicide in Queensland, Australia, using a spatiotemporal approach.

Methods

Seasonal data on suicide, demographic variables and socioeconomic indexes for areas in each Local Government Area (LGA) between 1999 and 2003 were acquired from the Australian Bureau of Statistics. Climate data were supplied by the Australian Bureau of Meteorology. A multivariable generalized estimating equation model was used to examine the impact of socio-environmental factors on suicide.

Results

The preliminary data analyses show that far north Queensland had the highest suicide incidence (e.g., Cook and Mornington Shires), while the southwestern areas had the lowest incidence (e.g., Barcoo and Bauhinia Shires) in all the seasons. Maximum temperature, unemployment rate, the proportion of Indigenous population and the proportion of population with low individual income were statistically significantly and positively associated with suicide. There were weaker but not significant associations for other variables.

Conclusion

Maximum temperature, the proportion of Indigenous population and unemployment rate appeared to be major determinants of suicide at a LGA level in Queensland.

Background

Suicide is one of the major causes of mortality around the world with about 877,000 suicide deaths each year globally [1]. Socio-environmental impacts on mental health, including suicide, have drawn increasing research attention, especially in recent years as global socio-environmental conditions change rapidly [2,3].

A number of studies have examined the impact of meteorological factors on suicide and found that lower suicide rates were associated with increased rainfall [4], decreased temperature [5], decreased humidity [6], and increased sunshine [7]. Additionally, some studies indicated that suicide rates varied with season [8,9]. Socioeconomic status [10,11], unemployment rate [12-14], country of birth [15,16], governmental policy [17,18] and intervention [19,20] were also associated with suicide in different countries and areas.

Most of the previous suicide studies have focused on either meteorological or socioeconomic factors alone, and none has examined their combined effect. As all these factors can influence suicide in different aspects, the impact of these factors on suicide, thus, should be studied in a systematic way, to help formulating effective suicide prevention strategies. In addition, few of the previous studies have applied geographical information system (GIS) and or spatial analysis approaches to assess the geographical difference of suicide, and the socio-environmental impact on suicide [21,22].

This study examined the association of socio-environmental determinants with suicide in Queensland, Australia using a GIS-based ecological study design. Queensland is the second largest states in Australia and it lies on the northeast of the continent, covering an area about 1.73 million km2, with a total population about 4.18 million in June 2007. Southeast Queensland (SEQ) accounts for less than 1.3% of total area, but had 65.4% of total population [23]. Other places, especially the inland areas, have much less population density than that of the whole state. The whole state is divided into a few regions as Figure 1. The climate conditions vary across the whole states. The far north and coastal areas are hot and humid in summer, while the highlands near coast and south-eastern coasts are warm and humid in summer. The inland areas in the southeast have temperate or warm summer, but cold winter. The western areas of Queensland have hot and dry summers, and mild and cold winters [24]. Queensland had more rapid increase in economy than the rest areas of Australia between 1992 until now, except for the financial year 1995-1996 [25]. The major industries in Queensland are agriculture, mining, financial services and tourism.

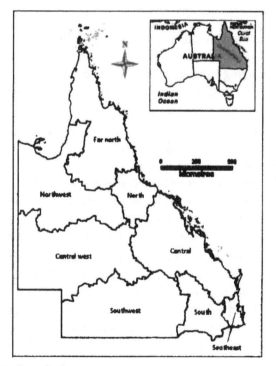

Figure 1. The regions in Queensland

Methods

Study Design

Spatiotemporal analysis of the impact of socio-environmental factors on suicide is critical because the distribution of suicide deaths and its determinants may vary with time and place, especially in Queensland, a large state with a wide range of climatic conditions and socioeconomic positions. The study consisted of four phases: data collection, data linkage and management, descriptive analyses, bivariable and multivariable analyses.

Data Sources

The meteorological data, including monthly rainfall (RF), maximum temperature (MaxT) and minimum temperature (MinT) were supplied by the Australian Bureau of Meteorology. Suicide, socioeconomic and demographic data were obtained from the Australian Bureau of Statistics (ABS).

The suicide data, covering a five-year period (1999-2003), included information on age, sex, year and month of suicide and Statistical Local Area (SLA) code. There are 125 Local Government Areas (LGAs) in Queensland in 2001 and each LGA has one or more SLAs. Suicide data were transferred into the LGA-based data, using the LGA codes from Australian Standard Geographical Classification (ASGC) [26]. Average suicide counts in total and by gender were calculated for each season at the LGA level (September, October and November for spring; December, January and February for summer; March, April and May for autumn; and June, July and August for winter).

The meteorological database was composed of monthly grid ($0.25° * 0.25°$, longitude and latitude; equivalent to the area of about 25 km*25 km) data. We used Vertical Mapper, a GIS tool, to transfer the meteorological data into the LGA data. Vertical Mapper was incorporated into the MapInfo, which was then used as a platform to perform the data link, data transfer and spatial display. After primary data retrieving and transferring, the structure of monthly meteorological data at the LGA level was established. The means of seasonal meteorological data at LGA level were calculated from monthly data.

Socio-economic Indexes for Area (SEIFA) and demographic data at the LGA level were based on CDATA 2001 of ABS, a database which provides information of 2001 Australian Census of Population and Housing, digital statistical boundaries, base map data and socio-economic data. We directly applied SEIFA and demographic data from the CDATA in the analysis.

SEIFA included four indices: the Index of Relative Socio-economic Advantage and Disadvantage (i.e., IRSAD, the higher IRSAD index, the higher socioeconomic position), the Index of Relative Socio-economic Disadvantage (i.e., IRSD, reflecting disadvantage such as low income and education level, high unemployment and unskilled occupations), the Index of Economic Resources (i.e., IER, reflecting the general level of availability to economic resources of residents and households) and the Index of Education and Occupation (i.e., IEO, reflecting the general educational level and occupational skills of people). All these indices were obtained from CDATA 2001.

Demographic variables included population, Indigenous population, unemployed population, population with low individual income (below AU$ 200 per week) and low education level (Year 9 and below). Using these numbers the following statistics were calculated: the proportion of Indigenous population (PIP), unemployment rate (UER), proportion of population with low individual income (PPLII) and proportion of population with low education level (PPLEL).

Data Analyses

A series of GIS and statistical methods were used to analyse these data. MapInfo (including Vertical Mapper incorporated) was used to explore the spatial patterns of socio-environmental variables and suicide.

Univariable analysis was applied to describe characteristics of each variable (suicide and socio-environmental factors). This step is important because it can show the pattern of distribution of each variable, and then select appropriate approaches for bivariable and multivariable analysis. Pearson correlations were applied for bivariable analysis after some non-normally-distributed data (suicide mortality rate, rainfall, IRSD, PIP and UER) were transformed into approximately normally-distributed values by logarithm transformation. The multicollinearity was tested for selecting variables for the multivariable modelling process. The multivariable generalized estimating equation (GEE) regression models with a Poisson link were developed to assess the possible impact of socio-environmental factors on suicide, after adjustment for the effects of potential confounders. The GEE model is well suited to analyse the repeated longitudinal data (e.g., climate data) [27]. This approach has also been used in other studies [28,29]. Spatial autocorrelation is defined as an auto-correlated association of a certain spatial variable with its spatial location, which means observations have similar values if they are close to each other in geographical aspect [30]. In this study, spatial autocorrelation test was applied to examine the variation of suicide between small areas. Semivariogram analysis was used to explore the spatial structure and spatial autocorrelation of suicide mortalities in Queensland, where semivariogram values were calculated on the basis of residuals. If there is spatial autocorrelation in model residuals, values are typically low and the semivariance increases with separation distance [30,31]. Statistical Package for the Social Sciences (SPSS) and S+ SpatialStats software were used for data analysis.

Results

Univariable Analysis

Table 1 demonstrates the distribution of suicide in Queensland between 1999 and 2003 by gender, year and month. There were 2,445 suicide cases in Queensland, with 1,957 males (80.0%) and 488 females (20.0%). There was no significant difference in monthly variation of suicide by gender.

Table 1. Suicide counts by year and month for both male (upper value) and females (lower value, italics)*

Years	Jan.	Feb.	Mar.	Apr.	May	Jun.	Jul.	Aug.	Sep.	Oct.	Nov.	Dec.
1999	22	32	27	30	25	24	38	47	31	40	30	32
	7	*7*	*3*	*6*	*9*	*3*	*12*	*10*	*5*	*10*	*7*	*10*
2000	37	43	42	28	37	37	26	34	29	51	36	44
	12	*7*	*8*	*9*	*14*	*10*	*9*	*9*	*11*	*10*	*7*	*9*
2001	43	28	36	36	33	31	31	33	29	27	24	38
	7	*6*	*15*	*7*	*9*	*8*	*4*	*13*	*7*	*11*	*2*	*12*
2002	43	31	41	33	31	32	38	32	29	36	34	41
	11	*9*	*8*	*6*	*8*	*10*	*7*	*10*	*7*	*8*	*8*	*8*
2003	37	30	24	21	11	17	26	44	35	38	31	11
	10	*9*	*3*	*10*	*5*	*5*	*11*	*6*	*8*	*6*	*7*	*3*
Total	182	164	170	148	137	141	159	190	153	192	155	166
	47	*38*	*37*	*38*	*45*	*36*	*43*	*48*	*38*	*45*	*31*	*42*

In the population of the whole Queensland, 36.4% of males and 34.4% of females were 24-year age and below. 29.5% of males and 30.1% of females aged between 25 and 44. 23.5% of male population and 22.9% of female population were between 45 and 64-year-age. Other people aged at 65-year and above. In the age structure of suicides, 16.4% of males and 15.2% females were adolescents and youth (aged at 24-year or below). 46.9% of males and 48.6% of females aged between 25 and 44-year. 24.6% of male suicides and 25.8% of female suicides were between 45 and 64-year age. 12.1% of males and 10.4% of females were older-aged adults (65 and over).

Table 2 shows the characteristics of each variable (suicide mortality rate and socio-environmental variables) at the LGA level over seasons. It shows a range of variation for each variable. Some data are normally-distributed (MaxT, MinT, IRSAD, IER, IEO, PPLII in total and by gender, PPLEL in total and by gender).

Some other variables are non-normally-distributed (suicide mortality rate, suicide ASM by gender, rainfall, IRSD, PIP in total and by gender and UER in total and by gender).

Table 2. Characteristics of suicide mortality, socio-demographic and environmental factors*

	Mean	SD	Minimum	Percentiles			Maximum
				25	50	75	
Total mortality (per 100,000)	4.28	15.038	0.00	0.00	0.00	3.17	211.64
Male ASM rate (per 100,000)	6.96	25.872	0.00	0.00	0.00	3.94	410.68
Female ASM rate (per 100,000)	1.34	10.053	0.00	0.00	0.00	0.00	225.73
RF (mm)	195.0	201.53	0.1	76.5	143.8	237.8	1865.4
MinT (°C)	15.1	5.08	2.2	12.1	15.6	18.7	26.0
MaxT (°C)	28.0	4.36	16.7	25.0	28.2	30.9	39.1
IRSAD	935.61	41.476	831.36	910.32	930.64	962.72	1059.84
IRSD	957.71	69.305	472.08	946.32	972.48	992.40	1048.88
IER	942.25	52.481	835.52	903.68	939.36	975.44	1083.76
IEO	929.05	35.566	815.68	909.60	925.84	945.52	1064.32
Total PIP (%)	7.79	14.392	0.00	1.92	2.86	6.15	87.51
Male PIP (%)	7.43	13.938	0.00	1.88	2.98	5.73	86.65
Female PIP (%)	8.23	14.987	0.00	1.89	3.14	6.61	88.67
Total UER (%)	6.76	3.819	0.00	4.03	6.04	8.76	23.25
Male UER (%)	7.14	4.526	0.00	4.06	6.45	9.36	26.71
Female UER (%)	6.25	3.110	0.00	4.16	6.11	7.93	18.37
Total PPLII (%)	28.14	7.472	10.65	24.39	27.91	31.77	61.71
Male PPLII (%)	23.06	8.937	4.30	17.66	22.44	27.44	62.22
Female PPLII (%)	34.19	6.611	17.13	30.51	33.63	37.86	65.18
Total PPLEL (%)	22.75	5.468	11.91	19.21	22.94	26.10	47.41
Male PPLEL (%)	24.61	6.234	11.21	20.00	25.25	28.53	50.83
Female PPLEL (%)	20.64	4.992	9.01	17.25	20.70	23.27	43.91

*Note: ASM (age-standard mortality rates); RF (rainfall); MinT (minimum temperature); MaxT (max temperature); IRSAD (Index of Relative Socio-economic Advantage and Disadvantage); IRSD (Index of Relative Socio-economic Advantage and Disadvantage); IER (Index of Economic Resources); IEO (Index of Education and Occupation); PIP (proportion of indigenous population); UER (unemployment rate); PPLII (proportion of population with low individual income); PPLEL (proportion of population with low educational level).

Figures 2 to 5 demonstrated the spatial patterns of age-adjusted standard mortality (ASM) of male suicide in Queensland between different seasons. In spring, some of far north, northwest, south, some of southeast and central coast areas had higher suicide ASM, while the inland and south western areas had lower suicide ASM or no suicide record (Figure 2). During the summer time, far north, some of north west, southeast and coastal and some of the south areas had higher suicide ASM; southwest and some of the central south areas had lower suicide ASM or no suicide record (Figure 3). Figure 4 indicates the suicide ASM distribution in autumn. Far north, west, central, some of the coastal and southeast areas had higher suicide ASM; south, southwest and some of the central areas had lower suicide ASM or no suicide record. In winter, some of far north areas, northwest, some of the southeast and east areas had higher suicide mortality rate, while central, southwest and other areas had lower suicide mortality rate or even no suicide record (Figure 5).

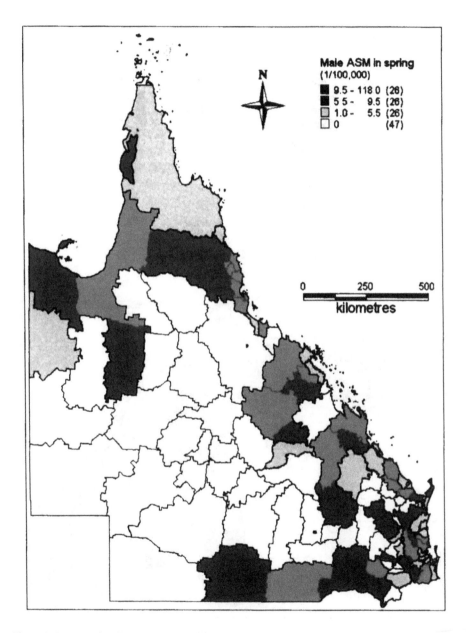

Figure 2. Average male ASM in spring (1999-2003)

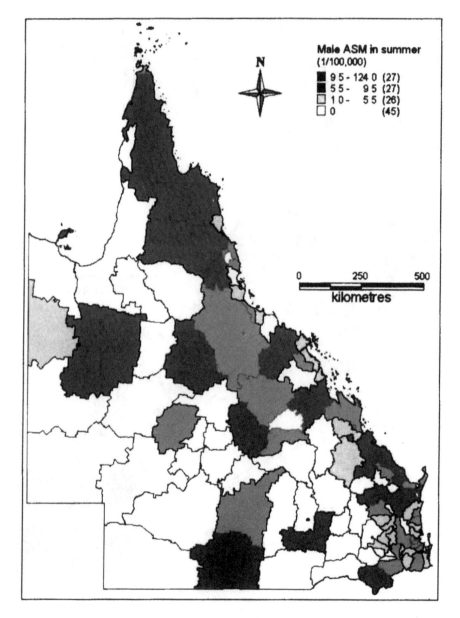

Figure 3. Average male ASM in summer (1999-2003)

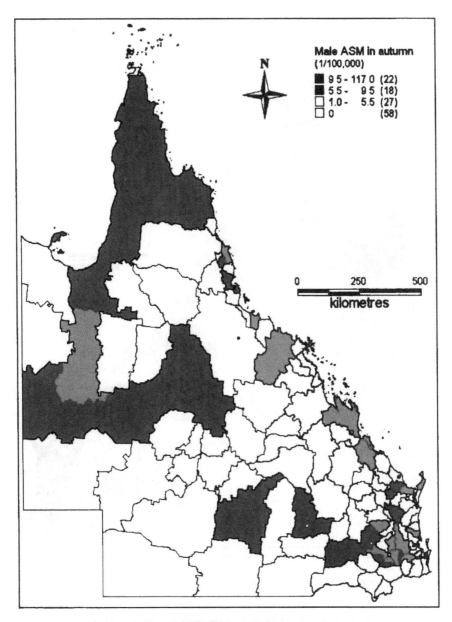

Figure 4. Average male ASM in autumn (1999-2003)

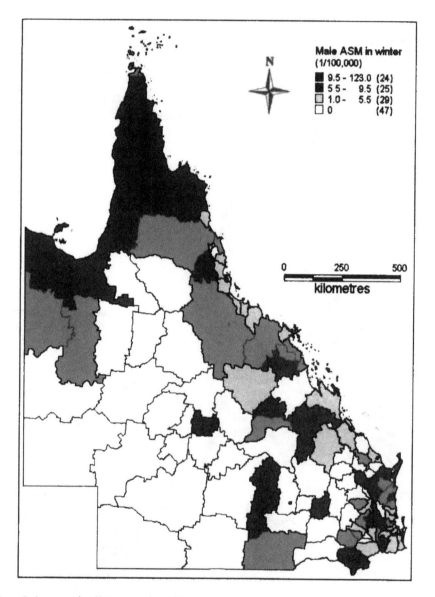

Figure 5. Average male ASM in winter (1999-2003)

The spatial patterns of ASM of female suicide across seasons were indicated in Figures 6 to 9. In spring, far north, north and central coast and some inland areas in the east and south had higher suicide ASM (Figure 6). There was higher suicide ASM in far north, some of north-western areas, some coastal and inland areas in the north and southeast than other areas (Figure 7). In autumn, some areas of central inland and coast, and southeast had higher suicide ASM compared with lower suicide ASM or no suicide record in other areas (Figure 8). There was higher suicide ASM in some parts of far north, central inland and south-eastern areas than other areas (Figure 9). 68.8% to 76.8% of LGAs had no female suicide record within each seasons.

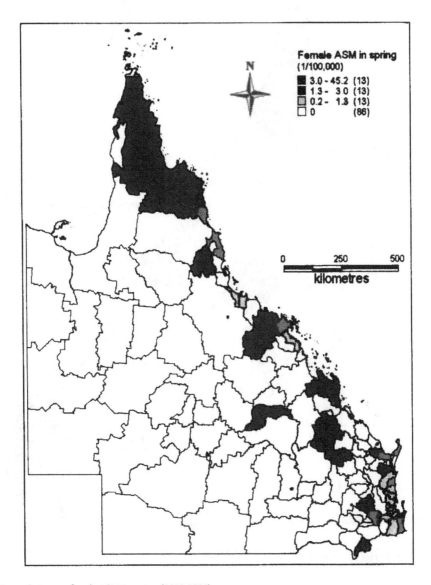

Figure 6. Average female ASM in spring (1999-2003)

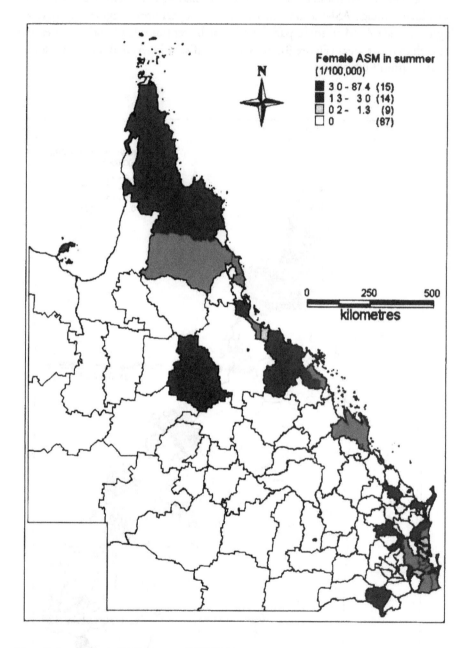

Figure 7. Average female ASM in summer (1999-2003)

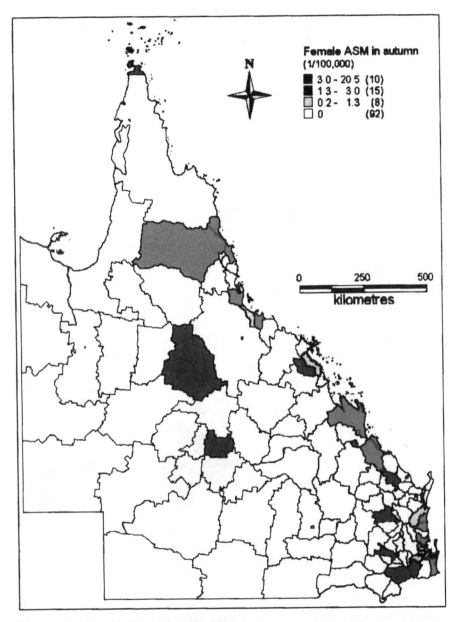

Figure 8. Average female ASM in autumn (1999-2003)

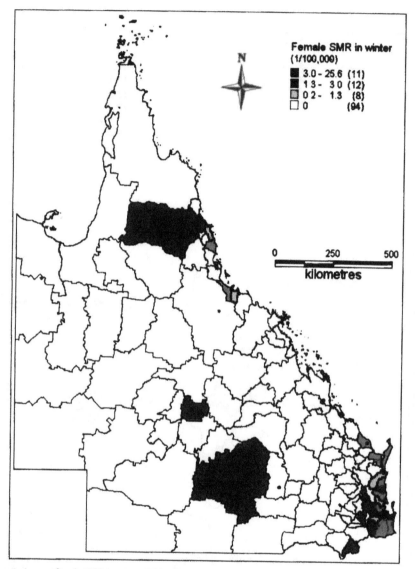

Figure 9. Average female ASM in winter (1999-2003)

In general, male suicides were recorded in the most LGAs, except for the southwest and some part of the central areas. Only 17 LGAs (13.6% of total) had no suicide recorded in the whole study period. Most female suicides occurred in the southeast, coastal and far north areas. There were very few female suicides in the majority of inland areas Queensland. Almost half (61 of 125) of total LGAs had no female suicide between 1999 and 2003.

Bivariable Analysis

Table 3 demonstrates the correlations between socio-environmental variables and suicide ASM by gender. In males, MinT, MaxT, PIP, PPLII and PPLEL were significantly and positively associated with suicide ASM, while there were negative associations between SEIFA and suicide. However, UER and RF were not significantly associated with male suicide. In females, RF was significantly and positively associated with suicide. SEIFA variables had significant and negative association with suicide except for IEO. PIP, PPLII and PPLEL were significantly and positively associated with suicide. UER ($p = 0.062$), MinT ($p = 0.088$) and IEO ($p = 0.087$) were marginally associated with suicide. However, MaxT ($p = 0.502$) was not significantly associated with suicide in females.

Table 3. Pearson correlations between socio-environmental factors and suicide for both male (upper value) and females (lower value, italics)

	ASM	RF	MinT	MaxT	IRSAD	IRSD	IER	IEO	PIP	UER	PPLII
RF	0.017 / *0.053**	1.000									
MinT	0.073** / *0.034*	0.449**	1.000								
MaxT	0.076** / *0.013*	0.163**	0.848**	1.000							
IRSAD	-0.147** / *-0.061**	-0.007	0.029	0.017	1.000						
IRSD	-0.296** / *-0.094**	-0.125**	-0.233**	-0.196**	0.606**	1.000					
IER	-0.091** / *-0.055**	-0.022	0.104**	0.108**	0.901**	0.361**	1.000				
IEO	-0.141** / *-0.034*	0.090**	-0.053**	-0.140**	0.772**	0.628**	0.445**	1.000			
PIP	0.163** / *0.069**	0.091** / *0.075**	0.257** / *0.250**	0.280** / *0.282**	-0.207** / *-0.189**	-0.603** / *-0.593**	-0.047* / *-0.015*	-0.277** / *-0.288**	1.000		
UER	0.001 / *0.037*	0.133** / *0.115**	-0.019 / *-0.018*	-0.221** / *-0.219**	-0.387** / *-0.310**	-0.150** / *-0.155**	-0.434** / *-0.314**	-0.067** / *-0.084**	-0.066** / *-0.107**	1.000	
PPLII	0.232** / *0.059**	0.120** / *0.015*	-0.019 / *-0.026*	-0.122** / *-0.067**	-0.677** / *-0.502**	-0.630** / *-0.516**	-0.625** / *-0.338**	-0.428** / *-0.561**	0.132** / *0.030*	0.493** / *0.251**	1.000
PPLEL	0.153** / *0.061**	-0.071** / *-0.012*	-0.003 / *0.009*	0.114** / *0.068**	-0.797** / *-0.786**	-0.615** / *-0.657**	-0.699** / *-0.672**	-0.672** / *-0.640**	0.357** / *0.376**	0.068** / *0.054**	0.557** / *0.478**

*Significant at the 0.05 level (2-tailed); **Significant at the 0.01 level (2-tailed).
†Note: ASM (age-standard mortality rates); RF (rainfall); MinT (minimum temperature); MaxT (max temperature); IRSAD (Index of Relative Socio-economic Advantage and Disadvantage); IRSD (Index of Relative Socio-economic Advantage and Disadvantage); IER (Index of Economic Resources); IEO (Index of Education and Occupation); PIP (proportion of Indigenous population); UER (unemployment rate); PPLII (proportion of population with low individual income); PPLEL (proportion of population with low educational level).

In the assessment of multicollinearity between socio-environmental variables, we found that some SEIFA indexes (e.g., IRSAD and IER) were highly correlated ($r = 0.90$). Thus, IRSD index was used to represent SEIFA in this study because of its strongest association with suicide across four SEIFA indexes. In addition, MinT and MaxT were also highly correlated ($r = 0.85$), and therefore, we use MaxT and MinT in separate models.

Multivariable Analysis

Multivariable GEE models were undertaken to examine the possible impact of climate variables, SEIFA and demographic factors on suicide mortality rate. We used the semivariance to measure the degree of spatial autocorrelation of model residuals. Figure 10 shows that there was no increased semivariance of residuals when the distance of LGAs increased. Thus it suggests that the GEE model fitted the data well as there was little spatial autocorrelation of residuals in this study.

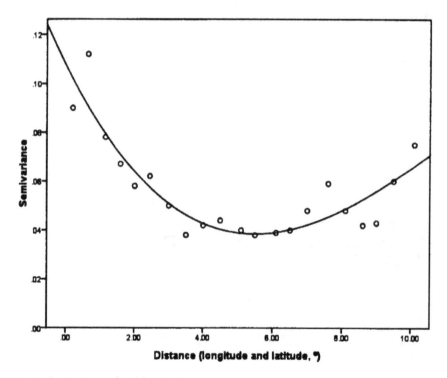

Figure 10. Semivariogram of model residuals

Table 4 shows the associations between socio-environmental variables and suicide mortality by gender. In males, MaxT, PIP and PPLII were significantly and positively associated with suicide in male population. RF, IRSD, UER and PPLEL were not significantly associated with male suicide. In females, PIP and UER were statistically significantly associated with suicide, but there was no significant association for other variables.

Table 4. Regression of socio-environmental determinants of suicide for both male (upper value) and females (lower value, italics)*

Variables	β	SE	RR (95% CI)	P-value
RF (mm)	0.090	0.073	1.09 (0.95 – 1.26)	0.220
	0.114	0.291	1.12(0.63 – 1.98)	0.696
MaxT (°C)	0.213	0.088	1.24 (1.04 – 1.47)	0.016
	-0.072	0.248	0.93(0.57 – 1.51)	0.773
IRSD	-0.017	0.026	0.98 (0.94 – 1.03)	0.483
	0.052	0.090	1.05 (0.88 – 1.26)	0.565
PIP (%)	0.065	0.030	1.07 (1.01 – 1.13)	0.029
	0.209	0.094	1.23 (1.03 – 1.48)	0.026
UER (%)	0.077	0.101	1.08 (0.89 – 1.32)	0.446
	0.087	0.042	1.09 (1.01 – 1.18)	0.036
PPLII (%)	0.373	0.086	1.45 (1.23 – 1.72)	0.000
	-0.176	0.137	0.84 (0.64 – 1.10)	0.198
PPLEL (%)	-0.139	0.117	0.87 (0.69 – 1.09)	0.234
	0.168	0.106	1.18 (0.96 – 1.46)	0.115

*Note: RR (relative risk); CI (confidential interval); RF (rainfall); MaxT (max temperature); IRSD (Index of Relative Socio-economic Advantage and Disadvantage); PIP (proportion of Indigenous population); UER (unemployment rate); PPLII (proportion of population with low individual income); PPLEL (proportion of population with low educational level).

Discussion

This study examined the relationship between socio-environmental factors and suicide using GIS and spatiotemporal analysis approaches. A range of climate, socioeconomic and demographic determinants were included in this quantitative analysis.

The results of this study indicate some key socio-environmental predictors of suicide at the LGA level. The preliminary spatiotemporal analyses show that far north Queensland had the highest suicide mortality, while the south-western areas had the lowest mortality rate in all the seasons. MaxT, PPLII and PIP were positively associated with total and male suicide. UER had a positive association with total and female suicide. RF had a significant and positive association with total suicide only. However, no significant association was found for SEIFA and PPLEL.

Some of the previous studies found that rainfall was negatively associated with suicide [4,32], while some other studies showed that this association was very weak [33,34]. Persistent rainfall deficiency results in drought, which causes reduction of crops in rural areas and adds financial burden to local residents, especially farmers [32,35]. In rural areas, farmers and other residents usually have less social support than urban residents, and this situation can get worse due to drought [36]. All these add stress, anxiety and mental health problems among the rural population which will eventually lead to suicidal behaviours and even suicide. However, this study only covered 5-year rainfall and suicide data, so it is difficult to determine the long term effect of rainfall on suicide. Another explanation for this discrepancy is that Queensland is in tropical and subtropical areas with much rainfall in general, especially in coastal areas. Even during drought periods, rainfall in Queensland is still much higher than other states in Australia.

In this study, higher MaxT was accompanied with increased suicide mortality at a LGA level. This finding corroborates previous reports [6,37,38]. For instance, some studies discovered that higher temperature can lead to decreased availability of tryptophan in human body, one of the 20 standard amino acids, then the volume of 5-Hydroxyindoleacetic acid (5-HIAA) synthesized from tryptophan greatly reduced [39]. As 5-HIAA can reduce depression among humans [40], therefore, the reduced 5-HIAA indirectly caused by high temperature leads to more depression and other mental health problems among population, even suicidal behaviours. We also examined the association between minimum temperature and suicide in the GEE model, but the association was very weak.

This study demonstrates a general trend that LGAs with higher PIP had higher rates of suicide. As most of the Indigenous population are located in rural areas, these communities often have lower SES and less opportunities of healthcare, including mental health services. The rapid social change in Australia may also affect the Indigenous communities, with more unhealthy behaviours such as excessive alcohol use and family violence [41]. The environmental injustice in this study should not be ignored. Some activities (e.g., construction of water systems, land use, and management of organizations) may cause cultural, environmental and economic risks and hazards among the local communities, especially in the areas with low SES and high proportion of Indigenous population [42,43]. The above factors contribute to the higher suicide occurrence and deaths in communities with a high proportion of Aboriginal and Torres Strait Islanders in Queensland [44]. Other studies in the United States also indicate that suicide mortality rates were higher in the areas with higher proportion of Indigenous population than in the other areas [45-47]. These studies also discovered that suicide is associated with harsh environmental and social conditions.

Increased unemployment rate directly reduces individual and family income, and thus can cause the financial burden and result in anxiety and stress among family members, especially for a less skilled population. These may increase the risk of mental health problems and suicidal behaviours. This can explain why unemployment had an adverse impact on suicide. Previous studies also discovered that higher unemployment rate can enhance the risk of suicide behaviour and suicide

[48-50]. In this study, unemployment had more significant impact on female suicide than male suicide. In recent years, more females participated in labour force than before, thus more females would experience unemployment as a consequence [51]. A study in Portugal also indicated that female suicide increased, as women play more important roles in the socioeconomic status and the stress of unemployment on them was more prominent than before [52].

Previous studies have indicated that higher socioeconomic status (SES) areas usually have lower suicide mortality [10,11], as high SES areas usually have higher employment rates, increased income and more accesses to training and education, compared with low SES areas [53]. In this study, we did not find a significant association between SEIFA and suicide, which may be due to a short time series dataset (5 years) and the use of a snapshot measure of SEIFA (i.e., disadvantage index in 2001).

Some studies indicate that the population with low income had higher suicide rate [54,55]. Generally, rural areas have higher proportion of population with low income, while healthcare (including mental health care) facilities are less developed and less accessible than urban areas. This can lead to increased mental health problems, even suicidal behaviours, among the local population. The results of this study are consistent with previous studies.

Some studies conducted in temperate areas like Brazil [56] and Italy [57] observed a peak of suicide in late spring and early summer. In this study, there were suicide peaks in August and October between 1999 and 2003. More suicides occurred in summer than other seasons. The results in this study were not completely consistent with previous studies, partly because all the LGAs of Queensland are in tropical and subtropical zones, and the four seasons are not evident in many places, especially in the north Queensland.

This study has several strengths. Firstly, this is the first study to examine an association between a wide range of socio-environmental factors and suicide at a LGA level in Queensland. Secondly, this study used a comprehensive spatial dataset, GIS and a range of quantitative analytical methods to compare the differences of socio-environmental impact on suicide over time and space. Thirdly, this study examined how socio-environmental factors influence the likelihood of suicide after taking into account a range of confounding factors, including gender, population size and SEIFA at the LGA level. Finally, the results of this study may have implications in public health policy making and implementation of suicide prevention intervention.

The limitations of this study should also be acknowledged. Firstly, the time series data set for analysis is short, compared with other studies [32,37,57]. Secondly, climate condition varies in different zones within each LGA, especially those covering large areas. So it is difficult to actually determine the climate condition in the geographical spot of each suicide death. Thirdly, the SEIFA index and demographic data at the LGA level were only based on 2001 Population Census, so it cannot reflect any changes in socioeconomic and demographic features during the whole study period. Thus the results of this study should be interpreted cautiously. Finally, this study only included several socio-environmental variables (i.e., rainfall, temperature, SEIFA, demographic variables), while other factors like personal and family history of mental health and psychiatric problems [58,59], local health service facilities [60], nutrition [61,62], religion [63,64], alcohol and drug use [65-67] may also influence mental health status and suicidal behaviours. However, the information on these variables was unavailable in this study.

The impact of socio-environmental change on mental health has drawn much attention. On the one hand, the current global financial crisis is likely to deepen, and it will almost certainly have negative effects on the trend of suicide. On the other hand, as climate change continues, the frequency, intensity and duration of weather extremes (e.g., flood, drought and cyclone) are likely to increase in the coming decade, it may also lead to the increase in suicide. Thus it is vital to strengthen surveillance system on weather extremes (e.g., high temperature) and social changes (e.g., unemployment) as well as the impacts of these changes on mental health [68-70]. Governmental officials, epidemiologists, psychiatrists, environmental health workers, economists, meteorologists and community leaders should work together to design, develop and implement effective suicide prevention and control strategies through an integrated and systematic approach.

Conclusion

In this study, we discovered that suicide ASM varied between LGAs by gender. Maximum temperature, the proportion of Indigenous population and unemployment rate appeared to be major determinants of suicide at a LGA level in Queensland. Other factors, such as rainfall, education and income level, had no significant association with suicide at a LGA level, during the period 1999-2003. These findings may have implications in planning and implementing population-based suicide interventions.

List of Abbreviations

ABS: (Australian Bureau of Statistics); ASM: (age-adjusted standardized mortality); CI: (confidential interval); GIS: (geographical information system); GEE: (generalized estimating equation); IRSAD: (Index of Relative Socio-economic

Advantage and Disadvantage); IRSD: (Index of Relative Socio-economic Advantage and Disadvantage); IEO: (Index of Education and Occupation); IER: (Index of Economic Resources); LGA: (Local Governmental area); MinT: (minimum temperature); MaxT: (max temperature); PIP: (proportion of Indigenous population); PPLII: (proportion of population with low individual income); PPLEL: (proportion of population with low educational level); RF: (rainfall); RR: (relative risk); SEIFA: (Socioeconomic Indexes for Areas); SLA: (statistical local area); UER: (unemployment rate).

Competing Interests

The authors declare that they have no competing interests.

Authors' Contributions

XQ designed the study, implemented all statistical analyses and drafted the manuscript. ST conceptualised the idea and revised the study protocol, especially the research design and data analysis. WH contributed to statistical analyses and interpretation of the results. All the authors contributed to the preparation of the final manuscript and approved the submission.

Acknowledgements

We owe much to Dr. Andrew Page of the University of Queensland, for providing the suicide data, Dr. Aaron Walker and Mr. Hang Jin from Queensland University of Technology (QUT) for valuable support in linking and arranging the climate data, and Dr. Adrian Barnett from QUT for contributing to data analysis and modeling.

References

1. World Health Organization: The world health report: shaping the future. Geneva, Switzerland; 2003.

2. Kefi S, van Baalen M, Rietkerk M, Loreau M: Evolution of local facilitation in arid ecosystems. American Naturalist 2008, 172:E1–17.

3. McCoy B: Suicide and desert men: the power and protection of kanyirninpa (holding). Australasian Psychiatry 2007, 15(Suppl):63–67.

4. Preti A, Miottob P: Seasonality in suicides: the influence of suicide method, gender and age on suicide distribution in Italy. Psychiatry Research 1998, 81:219–231.

5. Lin HC, Chen CS, Xirasagar S, Lee HC: Seasonality and climatic associations with violent and nonviolent suicide: a population-based study. Neuropsychobiology 2008, 57:32–37.

6. Deisenhammer EA, Kemmler G, Parson P: Association of meteorological factors with suicide. Acta Psychiatrica Scandinavica 2003, 108:455–459.

7. Linkowski P, Martin F, De Maertelaer V: Effect of some climatic factors on violent and non-violent suicides in Belgium. Journal of Affective Disorders 1992, 25:161–166.

8. Burns A, Goodall E, Moore T: A study of suicides in Londonderry, Northern Ireland, for the year period spanning 2000–2005. Journal of Forensic & Legal Medicine 2008, 15:148–157.

9. Rocchi MB, Sisti D, Miotto P, Preti A: Seasonality of suicide: relationship with the reason for suicide. Neuropsychobiology 2007, 56:86–92.

10. Page A, Morrell S, Taylor R, Carter G, Dudley M: Divergent trends in suicide by socio-economic status in Australia. Social Psychiatry & Psychiatric Epidemiology 2006, 41:911–917.

11. Taylor R, Page A, Morrell S, Harrisonb J, Carterc G: Mental health and socioeconomic variations in Australian suicide. Social Science & Medicine 2005, 61:1551–1559.

12. Morrell S, Taylor R, Kerr C: Unemployment and young people's health. Medical Journal of Australia 1998, 168:236–240.

13. Inoue K, Tanii H, Kaiya H, Abe S, Nishimura Y, Masaki M, Okazaki Y, Nata M, Fukunaga T: The correlation between unemployment and suicide rates in Japan between 1978 and 2004. Legal Medicine 2007, 9:139–142.

14. Iverson L, Andersen O, Andersen PK, Christoffersen K, Keiding N: Unemployment and mortality in Denmark, 1970–80. British Medical Journal 1987, 295:879–884.

15. Burvill PW: Migrant suicide rates in Australia and in country of birth. Psychological Medicine 1998, 28:201–208.

16. Westman J, Sundquist J, Johansson LM, Johansson SE, Sundquist K: Country of birth and suicide: a follow-up study of a national cohort in Sweden. Archives of Suicide Research 2006, 10:239–248.

17. Page A, Morrell S, Taylor R: Suicide and political regime in New South Wales and Australia during the 20th century. Journal of Epidemiology & Community Health 2002, 56:766–772.

18. Shaw M, Dorling D, Smith GD: Mortality and political climate: how suicide rates have risen during periods of Conservative government, 1901–2000. Journal of Epidemiology & Community Health 2002, 56:723–725.

19. Oravecz R, Czigler BA, Moore M: The transformation of suicide fluctuation in Slovenia. Archives of Suicide Research 2006, 10:69–76.

20. Hall WD, Mant A, Mitchell PB, Rendle VA, Hickie IB, McManus P: Association between antidepressant prescribing and suicide in Australia, 1991–2000: trend analysis. British Medical Journal 2003, 326:1008.

21. Middleton N, Sterne J, Gunnell DJ: An atlas of suicide mortality: England and Wales, 1988–1994. Health & Place 2008, 14:492–506.

22. Saunderson TR, Langford IH: A study of the geographical distribution of suicide rates in England and Wales 1989–92 using empirical Bayes estimates. Social Science & Medicine 1996, 43:489–502.

23. Australian Bureau of Statistics: Regional Population Growth, Australia and New Zealand, 1991 to 2001. Cat. No.3218.0. Canberra, Australia; 2002.

24. Australian Bureau of Meteorology: Australian climatic zones (based on temperature and humidity). [http://www.bom.gov.au/cgi-bin/climate/cgi_bin_scripts/clim_classification.cgi]

25. Queensland Government Treasury: 2007–08 Annual Economic Report on the Queensland economy - year ended 30 June 2008. Brisbane, Australia. 2008.

26. Australian Bureau of Statistics: Australian Standard Geographical Classification (ASGC) 2001. Cat. No.1216.0. Canberra, Australia; 2001.

27. Hanley JA, Negassa A, Edwardes MD, Forrester JE: Statistical analysis of correlated data using generalized estimating equations: An orientation. American Journal of Epidemiology 2003, 157:364–375.

28. Brooker S, Beasley M, Ndinaromtan M, Madjiouroum EM, Baboguel M, Djenguinabe E, Hay SI, Bundy DA: Use of remote sensing and a geographical information system in a national helminth control programme in Chad. Bulletin World Health Organization 2002, 80:783–789.

29. Chen XK, Yang Q, Smith G, Krewski D, Walker M, Wen SW: Environmental lead level and pregnancy-induced hypertension. Environmental Research 2006, 100:424–430.

30. Carl G, Kühn I: Analyzing spatial autocorrelation in species distribution using Gaussian and logit models. Ecological modelling 2007, 207:159–170.

31. Bell N, Schuurman N, Hameed SM: Are injuries spatially related? Join-count spatial autocorrelation for small-area injury analysis. Injury Prevention 2008, 14:346–353.

32. Nicholls N, Butler CD, Hanigan I: Inter-annual rainfall variations and suicide in New South Wales, Australia, 1964–2001. International Journal of Biometeorology 2006, 50:139–143.

33. Deisenhammer EA, Kemmler G, Parson P: Association of meteorological factors with suicide. Acta Psychiatrica Scandinavica 2003, 108:455–459.

34. Preti A: The influence of climate on suicidal behaviour in Italy. Psychiatry Research 1998, 78:9–19.

35. Sartore G: Drought and its effect on mental health–how GPs can help. Australian Family Physician 2007, 36:990–993.

36. Fuller J, Kelly B, Sartore G, Fragar L, Tonna A, Pollard G, Hazell T: Use of social network analysis to describe service links for farmers' mental health. Australian Journal of Rural Health 2007, 15:99–106.

37. Ajdacic-Gross V, Lauber C, Sansossio R, Bopp M, Eich D, Gostynski M, Gutzwiller F, Rossler W: Seasonal Associations between Weather Conditions and Suicide-Evidence against a Classic Hypothesis. American Journal of Epidemiology 2007, 165:561–569.

38. Vandentorren S, Bretin P, Zeghnoun A, Mandereau-Bruno L, Croisier A, Cochet C, Ribéron J, Siberan I, Declercq B, Ledrans M: August 2003 heat wave in France: risk factors for death of elderly people living at home. European Journal of Public Health 2006, 16:583–591.

39. Maes M, De Meyer F, Thompson P, Peeters D, Cosyns P: Synchronised annual rhythm in violent suicide rate, ambient temperature and light-dark span. Acta Psychiatrica Scandinavica 1994, 90:191–196.

40. Bell C, Abrams J, Nutt D: Tryptophan depletion and its implication for psychiatry. British Journal Psychiatry 2001, 178:399–405.

41. Measey MA, Li SQ, Parker R, Wang Z: Suicide in the Northern Territory, 1981–2002. Medical Journal of Australia 2006, 185:315–319.

42. Hillman M: Environmental justice: a crucial link between environmentalism and community development? Community Development Journal 2002, 37:349–360.

43. Hillman M: Situated justice in environmental decision-making: Lessons from river management in Southeastern Australia. Geoforum 2006, 37:695–707.

44. Hunter E, Milroy H: Aboriginal and Torres Strait islander suicide in context. Archives of Suicide 2006, 10:141–147.

45. Else IR, Andrade NN, Nahulu LB: Suicide and suicidal-related behaviors among indigenous Pacific Islanders in the United States. Death Studies 2007, 31:479–501.

46. Seale JP, Shellenberger S, Spence J: Alcohol problems in Alaska Natives: lessons from the Inuit. American Indian & Alaska Native Mental Health Research 2006, 13:1–31.

47. Wexler L, Hill R, Bertone-Johnson E, Fenaughty A: Correlates of Alaska Native fatal and nonfatal suicidal behaviors 1990–2001. Suicide & Life Threat Behavior 2008, 38:311–320.

48. Chan WS, Yip PS, Wong PW, Chen EY: Suicide and unemployment: what are the missing links? Archives of Suicide Research 2007, 11:327–335.

49. Fergusson DM, Boden JM, Horwood LJ: Recurrence of major depression in adolescence and early adulthood, and later mental health, educational and economic outcomes. British Journal of Psychiatry 2007, 191:335–342.

50. Yasan A, Danis R, Tamam L, Ozmen S, Ozkan M: Socio-cultural features and sex profile of the individuals with serious suicide attempts in southeastern Turkey: a one-year survey. Suicide & Life Threat Behavior 2008, 38:467–480.

51. Stack S: Suicide: a 15 year review of the sociological literature: Part I: Cultural and economic factors. Suicide and Life Threatening Behavior 2000, 30:163–176.

52. De Castro EF, Pimenta I, Martins I: Female independence in Portugal: effect on suicide rates. Acta Psychiatrica Scandinavica 1988, 78:147–155.

53. Australian Bureau of Statistics: Measures of Australia's progress: summary indicators, 2005. Cat No. 1383.0.55.001. Canberra, Australia. 2005.

54. Huisman M, Oldehinkel AJ: Income inequality, social capital and self-inflicted injury and violence-related mortality. Journal of Epidemiology & Community Health 2009, 63:31–37.

55. Kalist DE, Molinari NA, Siahaan F: Income, employment and suicidal behavior. Journal of Mental Health Policy & Economics 2007, 10:177–187.

56. Benedito-Silva AB, Nogueira Pires ML, Calil HM: Seasonal variation of suicide in Brazil. Chronobiology International 2007, 24:727–737.

57. Preti A, Lentini G, Maugeri M: Global warming possibly linked to an enhanced risk of suicide: Data from Italy, 1974–2003. Journal of Affective Disorders 2007, 102:19–25.

58. Kalmar S, Szanto K, Rihmer Z, Mazumdar S, Harrison K, Mann JJ: Antidepressant prescription and suicide rates: effect of age and gender. Suicide & Life Threatening Behavior 2008, 38:363–374.

59. Landberg J: Alcohol and suicide in Eastern Europe. Drug & Alcohol Review 2008, 27:361–373.

60. Pirkola S, Sund R, Sailas E, Wahlbeck K: Community mental-health services and suicide rate in Finland: a nationwide small-area analysis. Lancet 2009, 373:147–153.

61. Lakhan SE, Vieira KF: Nutritional therapies for mental disorders. Nutrition Journal 2008, 7:2.

62. Li Y, Zhang J, McKeown RE: Cross-sectional assessment of diet quality in individuals with a lifetime history of attempted suicide. Psychiatry Research 2009, 165:111–119.

63. Rasic DT, Belik SL, Elias B, Katz LY, Enns M, Sareen JZ: Spirituality, religion and suicidal behavior in a nationally representative sample. Journal of Affective Disorders 2009, 114:32–40.

64. Dervic K, Oquendo MA, Grunebaum MF, Ellis S, Burke AK, John Mann J: Religious affiliation and suicide attempt. American Journal of Psychiatry 2004, 161:2303–2308.

65. Evren C, Sar V, Evren B, Dalbudak E: Self-mutilation among male patients with alcohol dependency: the role of dissociation. Comprehensive Psychiatry 2008, 49:489–495.

66. Wang AG, Stórá T: Core features of suicide, gender, age, alcohol and other putative risk factors in a low-incidence population. Nordic Journal of Psychiatry 2008, 63:154–159.

67. Wojnar M, Ilgen MA, Czyz E, Strobbe S, Klimkiewicz A, Jakubczyk A, Glass J, Brower KJ: Impulsive and non-impulsive suicide attempts in patients treated for alcohol dependence. Journal of Affective Disorders 2008, 115:131–139.

68. Abaurrea J, Cebrián AC: Drought analysis based on a cluster Poisson model: distribution of the most severe drought. Climate Research 2002, 22:227–235.

69. Diaz JH: The influence of global warming on natural disasters and their public health outcomes. American Journal of Disaster Medicine 2007, 2:33–42.

70. Hoffpauir SA, Woodruff LA: Effective mental health response to catastrophic events: lessons learned from Hurricane Katrina. Family & Community Health 2008, 31:17–22.

Cycling and Walking to Work in New Zealand, 1991-2006: Regional and Individual Differences, and Pointers to Effective Interventions

Sandar Tin Tin, Alistair Woodward, Simon Thornley
and Shanthi Ameratunga

ABSTRACT

Background

Active commuting increases levels of physical activity and is more likely to be adopted and sustained than exercise programmes. Despite the potential health, environmental, social and economic benefits, cycling and walking are increasingly marginal modes of transport in many countries. This paper investigated regional and individual differences in cycling and walking to work in New Zealand over the 15-year period (1991-2006).

Methods

New Zealand Census data (collected every five years) were accessed to analyse self-reported information on the "main means of travel to work" from individuals aged 15 years and over who are usually resident and employed in New Zealand. This analysis investigated differences in patterns of active commuting to work stratified by region, age, gender and personal income.

Results

In 2006, over four-fifths of New Zealanders used a private vehicle, one in fourteen walked and one in forty cycled to work. Increased car use from 1991 to 2006 occurred at the expense of active means of travel as trends in public transport use remained unchanged during that period. Of the 16 regions defined at meshblock and area unit level, Auckland had the lowest prevalence of cycling and walking. In contrast to other regions, walking to work increased in Wellington and Nelson, two regions which have made substantial investments in local infrastructure to promote active transport. Nationally, cycling prevalence declined with age whereas a U-shaped trend was observed for walking. The numbers of younger people cycling to work and older people walking to work declined substantially from 1991 to 2006. Higher proportions of men compared with women cycled to work. The opposite was true for walking with an increasing trend observed in women aged under 30 years. Walking to work was less prevalent among people with higher income.

Conclusion

We observed a steady decline in cycling and walking to work from 1991 to 2006, with two regional exceptions. This together with the important differences in travel patterns by age, gender and personal income highlights opportunities to target and modify transport policies in order to promote active commuting.

Background

Physical activity provides substantial health benefits such as avoiding premature deaths [1], lowering the risk of a range of health conditions, notably cardiovascular diseases [2] and some forms of cancer [3], and enhancing emotional health [4]. While regular physical activity (i.e., undertaking at least 30 minutes of moderate intensity physical activity on most, if not all, days of the week) is recommended to promote and maintain health [5-7], maintenance of such activity has been identified as a major barrier for health behaviour interventions [8,9]. Previous research suggests that active commuting (building cycling and walking into daily life) may be more likely to be adopted and sustained compared with exercise programmes [10].

We have found published evidence of a variety of health benefits associated with active commuting. For example, obesity rates are lower in countries where active travel is more common [11]. A recent review reported that active commuting was associated with an 11% reduction in cardiovascular event rates [12]. A Copenhagen study found a 28% lower risk of mortality among those who cycled to work, even after adjusting for leisure time physical activity [13]. Similar associations were observed among Chinese women who cycled or walked for transportation [14]. In addition, active commuting may enhance social cohesion, community livability and transport equity [15-17], improve safety to all road users [18], save fuel and reduce motor vehicle emissions. A previous study predicted that if recommended daily exercise was swapped for transportation, this could reduce 38% of US oil consumption (for walking and cycling) and 11.9% of US's 1990 net emissions (for cycling), and could burn 12.2 kg of fat per person annually (for walking) and 26.0 kg of fat per person annually (for cycling) [19].

These effects are important not only in high-income countries in which the private motor vehicle has long been the dominant mode of transport but also in rapidly industrialising parts of the world, such as China, in which active commuting was until recently very common, but is now being replaced by motorised transport [20].

New Zealand is among the countries with the highest rate of car ownership in the world (607 cars per 1000 population) [21]. Driver or passenger trips account for four-fifths of the overall travel modal share [22] although one third of vehicle trips are less than two kilometres and two-thirds are less than six kilometres [23]. While the national Transport Strategy aims to "increase walking and cycling and other active modes to 30% of total trips in urban areas by 2040" [24], this target is unlikely to be met given current patterns of expenditure on the transport network [25].

Travel to work makes up about 15% of all travel in New Zealand [22]. Use of private motor vehicles is the dominant mode of travel to work [26] and may be sensitive to changing oil price [27]. The aim of this study was to investigate regional and individual differences in cycling and walking to work in the employed Census population over the 15-year period between 1991 and 2006. Possible intervention and policy options to promote active commuting will be discussed from New Zealand and international perspectives.

Methods

This paper presents an analysis of aggregate data obtained from the New Zealand Census undertaken by Statistics New Zealand every five years. Each Census since 1976 has collected information about the "main means of travel to work." However, the question was not date-specific prior to 1991.

The last four Censuses (1991, 1996, 2001 and 2006) asked usually resident employed persons aged 15 years and over about their main mode of transport to work on the date of Census (first Tuesday in March). For example, the 2006 Census asked the question "On Tuesday 7 March what was the one main way you travelled to work—that is, the one you used for the greatest distance?" and response options included: worked at home; did not go to work; public bus; train; drove a private car, truck or van; drove a company car, truck or van; passenger in a car, truck, van or company bus; motorbike; bicycle; walked or jogged; and other. The non-response rates to this particular question were 1.6%, 3.3%, 3.5% and 3.7% for the 1991, 1996, 2001 and 2006 Census respectively. The sample for this study was restricted to those who travelled to work on the specified day (i.e., those who reported "worked at home" or "did not go to work" were excluded, which ranged from 18% in 1991 to 22% in 2001).

The 'means of travel to work' responses were categorised into four main groups: "bicycle," "walk," "public transport" (including "public bus" and "train" responses) and "vehicle driver/passenger" (including "drove a private car, truck or van," "drove a company car, truck or van" and "passenger in a car, truck, van or company bus" responses). Trends in the main means of travel to work were presented for the 30-year period (1976 to 2006). As the data collected prior to 1991 were not date specific, the 1991 and 2006 Census data were used to examine trends in cycling and walking to work by region, age and gender. There are a total of 16 regions in New Zealand defined at meshblock and area unit levels: nine in the North Island and seven in the South Island. A meshblock is the smallest geographic area containing an average of 100 people and 40 dwellings [28]. Total personal income before tax in the 12 months ending 31 March was collected as a range and the data were analysed for the 2006 Census only due to limited comparability of data across Censuses. All data were self-reported and only aggregate data were available for this analysis.

The Ministry of Transport's Household Travel Survey data (2003-2008) [29] were used to compute the average distance of home to work trips in each region. It is a national survey collecting data on personal travel from about 3500 people (from about 2000 households) throughout New Zealand each year. The data were weighted to account for household and person non-responses. Information on other regional characteristics was obtained from the Statistics New Zealand (population density) [30] and the National Institute of Water & Atmospheric Research (climate status) [31]. The relationship between these characteristics and participation levels of active transport were measured using Spearman's rank correlation coefficient and linear and non-linear regression.

Results

The majority of people travelled to work by car, with an increasing trend over time from 64.8% in 1976 to 83.0% in 2006 (Figure 1). In contrast, walking to work declined over this 30 year period (12.8% in 1976 to 7.0% in 2006). The prevalence of cycling to work increased slightly from 1976 (3.4%) to 1986 (5.6%) and then declined steadily. In 2006, only 2.5% of people who travelled to work used a bicycle. The prevalence of public transport use decreased from 12.8% in 1976 to 5.1% in 1991 but remained stable at around 5.0% over the last 15 year period.

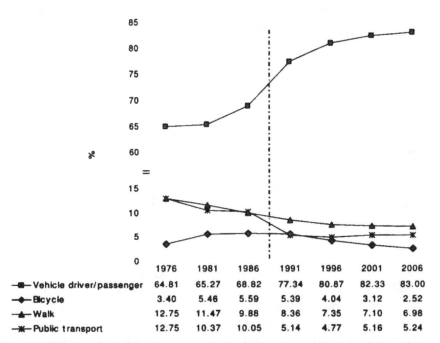

	1976	1981	1986	1991	1996	2001	2006
Vehicle driver/passenger	64.81	65.27	68.82	77.34	80.87	82.33	83.00
Bicycle	3.40	5.46	5.59	5.39	4.04	3.12	2.52
Walk	12.75	11.47	9.88	8.36	7.35	7.10	6.98
Public transport	12.75	10.37	10.05	5.14	4.77	5.16	5.24

Figure 1. Mode of travel to work on the census day in the usually resident employed population aged 15 years and over (1976 to 2006).

Regional Differences in Cycling and Walking to Work

Regional variation in active transport along with environmental and geographic factors thought to influence this variation is presented in Table 1. Auckland is the most populated region and West Coast, the least. The average distance of the trip to work varies from 6.7 km in West Coast to 14.8 km in Waikato. There is a moderate variation in average temperatures and sunshine hours with highest levels recorded in regions in the north of the South Island; and a three-fold varia-tion in rainfall across the major urban areas of different regions around the time of the census.

Table 1. Regional characteristics and correlations with the prevalence of cycling and walking to work

Region	Population density (per km²)[1] 2006	Average distance of home-work trips (km)[2] (95% CI) 2003-2006	Average sunshine (hours)[3] 1971-2000	Average rainfall (mm)[3] 1971-2000	Average air tem-perature (°C)[3] 1971-2000
Northland	10.8	12.2 (7.0-17.4)	153	144	18.6
Auckland	215.3	10.9 (9.9-12.0)	180	82	18.7
Waikato	15.9	14.8 (11.1-18.5)	184	87	17.1
Bay of Plenty	21.0	9.5 (6.8-12.1)	197	132	18.3
Gisborne	5.3	8.3 (5.2-11.5)	185	99	17.4
Hawke's Bay	10.5	9.2 (6.4-12.1)	194	85	17.7
Taranaki	14.3	9.3 (5.1-13.6)	202	108	16.9
Manawatu-Wanganui	10.0	9.5 (7.4-11.6)	170	74	16.6
Wellington	55.2	12.4 (10.2-14.6)	191	92	16.6
Tasman	4.6	8.7 (6.4-11.1)*	212	75	16.3
Nelson	96.8	8.7 (6.4-11.1)*	212	77	16.1
Marlborough	3.9	8.7 (6.4-11.1)*	224	54	16.3
West Coast	1.3	6.7 (5.5-7.9)	161	171	15.7
Canterbury	11.7	10.1 (7.6-12.6)	183	56	15.1
Otago	6.2	9.3 (6.1-12.6)	139	70	13.7
Southland	2.8	9.9 (6.6-13.3)	136	94	12.5
Spearman Correlation Coefficient (p-value)					
% cycling to work (2006)	-0.25 (0.4)	-0.64 (0.007)	0.58 (0.02)	-0.46 (0.07)	-0.36 (0.2)
% walking to work (2006)	-0.29 (0.3)	-0.27 (0.3)	-0.03 (0.9)	-0.15 (0.6)	-0.62 (0.01)

1 – Source: Indicator 2: Living density. Statistics New Zealand
2 – Source: Household Travel Survey data. Ministry of Transport
3 – Historical averages for the main cities/centres in March. Source: NIWA National Climate Database. National Institute of Water & Atmospheric Research
* - Average distance for three regions

Active travel to work varied widely across regions. In 2006, Nelson had the highest prevalence of cycling (7.2%) and Auckland, the lowest (1.0%) (Figure 2). All regions experienced a sharp fall in cycling prevalence, most steeply in Gisborne, over the 15 year period between 1991 and 2006. Walking prevalence was highest in Otago (11.3%), Wellington (11.1%) and West Coast (10.9%) and lowest in Auckland (4.9%). Contrary to other regional trends, the proportion of people who walked to work in Wellington and Nelson increased from 1991 to 2006.

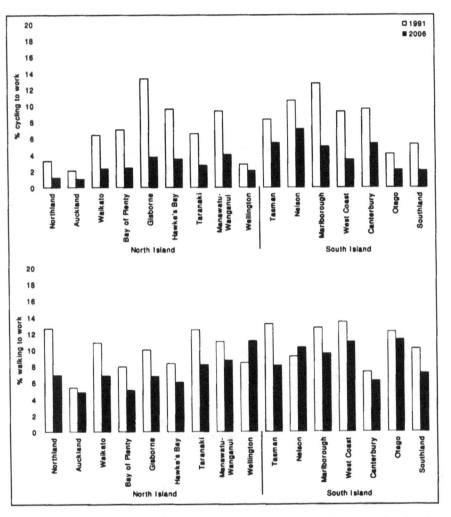

Figure 2. Proportion of people who cycled and walked to work on the census day by area of usual residence (1991 to 2006).

The prevalence of cycling to work was negatively correlated with the average distance of home to work trips and positively correlated with average sunshine hours whereas the prevalence of walking was negatively correlated with average air temperature ($p < 0.05$) (Table 1). Further explorations revealed the relationship between cycling prevalence and average distance to work to be log-linear and the relationships between cycling prevalence and average sunshine hours as well as walking prevalence and average temperature to be linear (Figure 3).

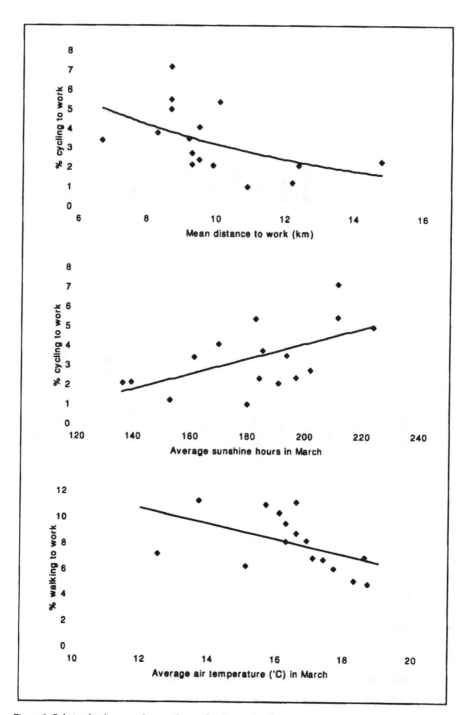

Figure 3. Relationship between the prevalence of cycling and walking to work and specific regional factors.

Individual Differences in Cycling and Walking to Work

Higher proportions of men compared with women cycled, while higher proportions of women walked to work (Figure 4). In 1991, the prevalence of cycling to work declined with age but this trend was less pronounced in 2006. The largest decline in cycling over the 15 year period was among younger age groups, particularly 15-19 year olds. Walking to work was least prevalent among middle-aged men and women. A significantly higher proportion of 15-29 year old women walked to work in 2006, compared with 1991. The prevalence of cycling to work did not vary significantly by personal income level whereas walking to work was less prevalent among people with higher income in 2006 (Figure 5).

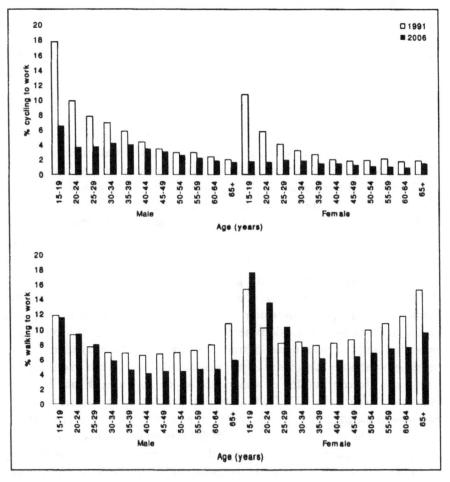

Figure 4. Proportion of people who cycled and walked to work on the census day by age and gender (1991 to 2006).

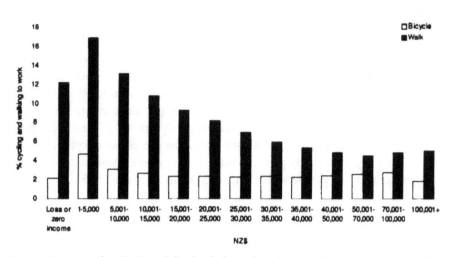

Figure 5. Proportion of people who cycled and walked to work on the census day by personal income (2006).

Discussion

Our analysis showed that more than four-fifths of New Zealanders used a private motor vehicle to travel to work on Census day in 2006. Only one in fourteen people walked to work and one in forty cycled. Increased car use from 1991 to 2006 occurred at the expense of active means of travel as the prevalence of using public transport remained unchanged during that period. We found important differences in active travel patterns by region, age, gender and personal income.

This is one of very few papers reporting population-based active travel behaviour in New Zealand. One of the major benefits of using Census data is that it is a near-complete survey of the general population (96.3% response rate in 2006) and the people's transport activity nationally, regionally and across different population subgroups over time may be compared. When interpreting these results, however, some limitations need to be considered. First, the Census question asked only for 'main means of travel to work' and did not take into account multiple transport modes, for example, walking and taking a bus in one journey. This means the contribution of walking to the journey to work may be underestimated. Second, the 1991-2006 Census questions were date-specific and the data may be biased seasonally, although the timing of Census day has been similar year to year. People's active transport activity may be overestimated in this case as the Census is usually in March when the weather is warm and relatively dry. Third, we were not able to adjust for potential confounders as only aggregate data were available for this analysis. For example, personal income may be related to an individual's age, gender and residential area, all of which independently, influence choice of travel to work. Finally, the findings may be affected by the "ecological fallacy" as averaged aggregate data were used to infer relationships, for example, between various regional characteristics (such as average distance to work) and the proportion of cycling and walking to work. These questions may be addressed in future studies which obtain individual level data.

Despite these limitations, our findings are consistent with and extend the evidence gained from previous research. Parallel to decreasing trends in active travel to work behaviour, overall travel mode share for cycling and walking has been declining steadily in New Zealand (from 4% and 21% respectively in 1989 to 1% and 16% respectively in 2006) [32]. During the same period, the annual distance driven in light 4-wheeled vehicles has been increasing—particularly among the 45-64 age group [33]. From 1990 to 2006, total greenhouse gas emissions increased by 25.7%, and emissions from road transport increased disproportionately (by 66.9%) [34]. In 2006, transport accounted for 42% of total emissions from the energy sector [35]. A recent report indicates that the air quality in Auckland is worsening due to emissions from increasing use of motor vehicles [36].

A study from the US shows that CO2 emissions from the transport sector will continue to rise unless vehicle kilometres travelled can be substantially reduced, as present trends in car use will overwhelm the gains that may result from technological advances such as changes in fuel type (e.g., biodiesel fuel) and motor vehicle efficiency (e.g., hybrid cars) [37]. The findings are unlikely to be different in the New Zealand context given the country's dispersed population (4.3 million people spread over 268,680 km²), low density cities and automobile centred transportation system.

Other studies have found that New Zealanders rarely cycle or walk even when travelling short distances. Walking represents only 39% of all trips under two kilometres and cycling accounts for three percent of all trips under two kilometres and two percent of all trips between two and five kilometres in the 2004-2007 household travel surveys [38]. Only one-fifth of New Zealanders surveyed in 2003 strongly endorsed plans to replace car trips with active modes such as cycling and walking on at least two days per week and less than half of the latter considered cycling for short distances [39,40]. Although a variety of factors can influence public attitudes and behaviour [41], these findings are likely to reflect decades of under-investment in public transport and cycling and walking infrastructure. In Auckland, the construction of motorways has been favoured consistently over alternative modes in transport planning over the past 50 years [42].

We observed regional differences in patterns of cycling and walking to work. Such differences may be partly explained by aspects of the physical environment such as weather, climate and topography (hilliness) [43-45] and distance to work [46]. The influence of environmental factors such as average temperatures and rainfall, however, should not be over-emphasized. A number of cities in North America and Europe have reported substantial increases in the prevalence of walking and cycling in the last decade, for example, daily ridership doubled in New York between 2001 and 2006 [47], yet have climates much less favourable than those of most parts of New Zealand.

We found low rates of cycling to work in regions with long average distances to work (≥ 10 km). Statistics New Zealand reported that on the Census day in 2006, 83% of people who walked to work travelled less than 5 km and 89% of those who cycled to work travelled less than 10 km [26]. Although distance to work is not easily changed, increased housing density, availability of public transport and investment in active transport infrastructure such as bicycle lanes and shared paths may improve engagement in active travel modes.

Two New Zealand regions that bucked the overall trends by revealing increasing levels of walking warrant further comment. Regional strategies in Wellington and Nelson have made substantial investments in active transport. Wellington has proposed an urban development strategy [48], based on the idea of a "growth spine" (a strip of land along which more intensive urban development is encouraged), a bus lane programme [49] and school, workplace and community travel plans [50]. In Nelson, pedestrian, cycling and urban growth strategies have been implemented with integration between transport planning and urban development teams [51]. Future research will be required to investigate the effectiveness of these and other active transport strategies being implemented.

Studies from other automobile dependent countries such as the US, UK and Australia have also reported a comparatively low level of cycling and walking to work [52-56], with important sociodemographic variations in the patterns of active travel. In general, men are more likely to cycle than women; and women are more likely to walk than men. Younger people are more likely to walk and cycle compared with older age groups. This is important because it will be necessary to boost walking and cycling rates in the older age groups to realise the potential health benefits of active transport. The cardio-protective effects of exercise relate much more closely to current activity than to past exposures [57]. Our study shows that walking is more common in lower income groups, whereas socioeconomic status does not appear to influence cycling. Similar patterns were observed in previous research [53,58,59]; however, others reported associations between cycling and income [44,60].

In contrast, in most European countries, walking and cycling make up at least one-fourth of all urban trips (45% in the Netherlands) and active travel patterns are universal across different segments of society; walking increases with age, cycling declines only slightly, women cycle as much as men, and people from all income classes cycle [11,58,61-64]. The success of European countries in promoting cycling and walking is attributed to the "coordinated implementation of the multi-faceted, mutually reinforcing set of policies" in the past few decades [58,61]. Components include: provision of better facilities for pedestrians and cyclists, extensive traffic calming of residential neighbourhoods, increased traffic regulation and enforcement, people oriented urban design, integration with public transport, comprehensive traffic education and training, restrictions on car ownership, use and parking [58,61] and workplace travel plans [65].

In countries like New Zealand, significant barriers exist to implementing such comprehensive measures to promote active commuting but much could be achieved in the short term. For example, although Australia has sprawling cities and a high rate of car ownership, the prevalence of cycling to work has increased substantially in some states in the last decade together with growing investments in bicycle infrastructure (for example, there was a 43% increase from 2001 to 2006 in cycle commuters in Melbourne) [66-68]. Likewise, we found an increasing trend of walking to work in the two New Zealand regions that have invested in sustainable transport strategies.

As an important initial step at the national level, a project has begun to build a cycleway network running the length of New Zealand [69]. While primarily intended to enhance tourism, the initiative has the potential to promote active commuting if a comprehensive cycle network plan is incorporated to strengthen connections between residential areas and key activity centres in urban and rural New Zealand. A potential way to move toward more attractive environments for active commuting without major infrastructural change is reducing the speed limit in residential streets, which currently is 50 km/hr in New Zealand compared with 30 km/hr (or less) in European countries [70]. Given a favourable trend in cycling and walking as a recreational activity in New Zealand [71], another useful step would be offering interventions promoting a modal shift, i.e., from using cars to walking and cycling, tailored to recreational cyclists and walkers. The effectiveness of such targeted behaviour change programmes has been reported in a previous review [72]. An Australian study showed that participants in mass cycling events, particularly novice riders and first-time participants, cycled more frequently in the month after the event [73].

Conclusion

Walking and cycling are increasingly marginal modes of travel to work in New Zealand and socio-demographic differences exist in such behaviour patterns. Increased walking to work, recorded in some regions, indicates that potential gains may be made with systematic promotion of active modes. Translating such successes to the national level requires political will and public support to redress decades of land use and transport policies that have prioritised car use. The rising cost of fuel linked with a resurgence in recreational cycling and walking may provide the impetus to seriously promote cycling and walking as safe and attractive choices for travel to work.

Competing Interests

The authors declare that they have no competing interests.

Authors' Contributions

STT has contributed to acquisition, analysis and interpretation of data and drafting the manuscript. AW, ST and SA have contributed to interpretation of data and revising the manuscript critically. All authors have given final approval of the version to be published.

Authors' Information

STT - MBBS, MPH, Assistant Research Fellow, Section of Epidemiology and Biostatistics, School of Population Health, University of Auckland, New Zealand

AW - MBBS, PhD, FAFPHM, Professor and Head of School, School of Population Health, University of Auckland, New Zealand

ST - MBChB, MPH, Assistant Research Fellow, Section of Epidemiology and Biostatistics, School of Population Health, University of Auckland, New Zealand

SA - MBChB, PhD, FAFPHM, Associate Professor, Section of Epidemiology and Biostatistics, School of Population Health, University of Auckland, New Zealand

Acknowledgements

We thank Mr Wayne Gough at the Statistics New Zealand for providing Census data on travel to work behaviour, Ms Lynley Povey at the Ministry of Transport for providing Household Travel Survey data, and Dr Andrew Tait and Ms Georgina Griffiths at the National Institute of Water &Atmospheric Research for providing climate data.

References

1. Lee I, Skerett P: Physical activity and all-cause mortality: what is the dose-response relation? Med Sci Sports Exerc 2001, 33:S459–471.

2. Berlin JA, Colditz GA: A meta-analysis of physical activity in the prevention of coronary heart disease. Am J Epidemiol 1990, 132:612–628.

3. Colditz GA, Cannuscio CC, Frazier AL: Physical activity and reduced risk of colon cancer: implications for prevention. Cancer Causes Control 1997, 8:649–667.

4. Biddle S: Emotion, mood and physical activity. In Physical activity and psychological well-being. Edited by: Biddle S, Fox K, Boutcher S. London: Routledge; 2000:63–87.

5. Ministry of Health: DHB Toolkit: Physical Activity. Wellington: Ministry of Health; 2003.

6. Haskell WL, Lee IM, Pate RR, Powell KE, Blair SN, Franklin BA, Macera CA, Heath GW, Thompson PD, Bauman A: Physical Activity and Public Health: Updated Recommendation for Adults from the American College of Sports Medicine and the American Heart Association. Med Sci Sports Exerc 2007, 39:1423–1434.

7. Strong WB, Malina RM, Blimkie CJR, Daniels SR, Dishman RK, Gutin B, Hergenroeder AC, Must A, Nixon PA, Pivarnik JM, Rowland T, Trost S, Trudeau F: Evidence Based Physical Activity for School-age Youth. J Pediatr 2005, 146:732–737.

8. Orleans CT: Promoting the maintenance of health behavior change: recommendations for the next generation of research and practice. Health Psychol 2000, 19:76–83.

9. Marcus BH, Dubbert PM, Forsyth LH, McKenzie TL, Stone EJ, Dunn AL, Blair SN: Physical Activity Behavior Change: Issues in Adoption and Maintenance. Health Psychol 2000, 19(1 Supplement):32–41.

10. Hillsdon M, Thorogood M, Anstiss T, Morris J: Randomised controlled trials of physical activity promotion in free living populations: a review. J Epidemiol Community Health 1995, 49:448–453.

11. Bassett DR Jr, Pucher J, Buehler R: Walking, cycling, and obesity rates in Europe, North America, and Australia. J Phys Act Health 2008, 5:795–814.

12. Hamer M, Chida Y: Active commuting and cardiovascular risk: A meta-analytic review. Prev Med 2008, 46:9–13.

13. Andersen LB, Schnohr P, Schroll M, Hein HO: All-cause mortality associated with physical activity during leisure time, work, sports, and cycling to work. Arch Intern Med 2000, 160:1621–1628.

14. Matthews CE, Jurj AL, Shu X-O, Li H-L, Yang G, Li Q, Gao Y-T, Zheng W: Influence of exercise, walking, cycling, and overall nonexercise physical activity on mortality in Chinese women. Am J Epidemiol 2007, 165:1343–1350.

15. Appleyard D: Livable Streets. Berkeley: University of California Press; 1981.

16. Community Livability: Helping to create attractive, safe, cohesive communities, [http://www.vtpi.org/tdm/tdm97.htm].

17. Litman T: Evaluating transportation equity: Guidance for incorporating distributional impacts in transportation planning. Victoria, BC: Victoria Transport Policy Institute; 2007.

18. Wittink R: Planning for cycling supports road safety. In Sustainable transport: Planning for walking and cycling in urban environments. Edited by: Tolley R. Cambridge: Woodhead Publishing Limited; 2003:172–188.

19. Higgins PAT: Exercise-based transportation reduces oil dependence, carbon emissions and obesity. Environ Conserv 2005, 32:197–202.

20. Wu Y: Overweight and obesity in China. BMJ 2006, 333:362–363.

21. The Economist: Pocket World in Figures 2009 Edition. London: The Economist; 2008.

22. Ministry of Transport: Comparing travel modes. Household Travel Survey v1.4 revised Jan 2008. Wellington: Ministry of Transport; 2008.

23. Ministry of Transport: New Zealand Transport Strategy. Wellington: Ministry of Transport; 2002.

24. Ministry of Transport: The New Zealand Transport Strategy. Wellington: Ministry of Transport; 2008.

25. Transport funding realigned and increased, [http://www.beehive.govt.nz/release/transport+funding+realigned+and+increased].

26. Statistics New Zealand: Commuting Patterns in New Zealand: 1996–2006. Wellington: Statistics New Zealand; 2009.

27. Badland HM, Duncan MJ, Schofield GM: Using Census data to travel through time in New Zealand: patterns in journey to work data 1981–2006. New Zealand Medical Journal 2009, 122:15–20.

28. Statistics New Zealand: Meshblock Database. [http:/ / www.stats.govt.nz/ NR/ rdonlyres/ DDC36215-A350-4FBE-AC71-2FB7EEB8836 1/ 0/ 2001Cenus-MeshblockDatabase.pdf], 2002.

29. New Zealand Household travel Survey, [http://www.transport.govt.nz/research/TravelSurvey/].

30. Statistics New Zealand: Indicator 2: Living Density. [http:/ / www.stats.govt.nz/ Publications/ StandardOfLiving/ housing-indicators/ 2-living-density.aspx]

31. NIWA National Climate Database, [http://cliflo.niwa.co.nz/].

32. Ministry of Transport: Sustainable and safe land transport: trends and indicators. Wellington: Ministry of Transport; 2007.

33. Ministry of Transport: Driver travel in cars, vans, utes and SUVs. Household Travel Survey v1.2. Revised May 2007. Wellington: Ministry of Transport; 2007.

34. Ministry for the Environment: New Zealand's Greenhouse Gas Inventory 1990–2006. Wellington: Ministry for the Environment; 2008.

35. Ministry of Economic Development: New Zealand Energy Greenhouse Gas Emissions 1990–2006. Wellington: Ministry of Economic Development; 2007.

36. Auckland City Council: State of the environment. Update 2007/2008. Auckland: Auckland City Council; 2009.

37. Ewing R, Bartholomew K, Winkelman S, Walters J, Chen D: Growing cooler: The evidence on urban development and climate change. Washington, D.C: The Urban Land Institute; 2008.

38. O'Fallen C, Sullivan C: Trends in trip chaining and tours: Analysing changes in New Zealanders' travel patterns using the ongoing New Zealand Household Travel Survey. NZ Transport Agency Research Report 373. Wellington: New Zealand Transport Agency; 2009.

39. Badland H, Schofield G: Perceptions of replacing car journeys with non-motorized travel: exploring relationships in a cross-sectional adult population sample. Prev Med 2006, 43:222–225.

40. Sullivan C, O'Fallon C: Increasing cycling and walking: an analysis of readiness to change. Land Transport New Zealand Research Report 294. Wellington: Land Transport New Zealand; 2006.

41. Cleland B, Walton D: Why don't people walk and cycle? Central Laboratories Report No: 528007.00. Wellington: Land Transport New Zealand; 2004.

42. Mees P, Dodson J: Backtracking Auckland: Bureaucratic rationality and public preferences in transport planning. Urban Research Program Issue Paper 5. Brisbane: Griffith University; 2006.

43. Nankervis M: The effect of weather and climate on bicycle commuting. Transp Res Part A: Policy and Practice 1999, 33:417–431.

44. Winters M, Friesen MC, Koehoorn M, Teschke K: Utilitarian bicycling: a multilevel analysis of climate and personal influences. Am J Prev Med 2007, 32:52–58.

45. Parkin J, Wardman M, Page M: Estimation of the determinants of bicycle mode share for the journey to work using census data. Transportation 2008, 35:93–109.

46. Badland HM, Schofield GM, Garrett N: Travel behavior and objectively measured urban design variables: associations for adults traveling to work. Health Place 2008, 14:85–95.

47. Safety in numbers, [http:/ / www.transalt.org/ files/ newsroom/ streetbeat/ 2009/ June/ 0604.html#safety_in_numbers].

48. Wellington City Council: Urban Development Strategy: Directing growth and delivering quality. Wellington: Wellington City Council; 2006.

49. Bus Lanes, [http://www.wellington.govt.nz/projects/ongoing/buslanes.html].

50. Regional travel plans programme, [http://www.gw.govt.nz/section2271.cfm].

51. Strategies and Plans, [http://www.nelsoncitycouncil.co.nz/thecouncil/strategies-plans.htm].

52. Plaut PO: Non-motorized commuting in the US. Transp Res Part D: Transport and Environment 2005, 10:347–356.

53. Pucher J, Renne JL: Socioeconomics of urban travel: Evidence from the 2001 NHTS. Transp Q 2003, 57:49–77.

54. Department for Statistics: Travel to work: Personal travel fact sheet - July 2007. London: Department for Transport; 2007.

55. Bell AC, Garrard J, Swinburn BA: Active transport to work in Australia: is it all downhill from here? Asia Pac J Public Health 2006, 18:62–68.

56. Mees P, O'Connell G, Stone J: Travel to Work in Australian Capital Cities, 1976–2006. Urban Policy Res 2008, 26:363–378.

57. Lennon SL, Quindry J, Hamilton KL, French J, Staib J, Mehta JL, Powers SK: Loss of exercise-induced cardioprotection after cessation of exercise. J Appl Physiol 2004, 96:1299–1305.

58. Pucher J, Buehler R: Making cycling irresistable: lessons from the Netherlands, Denmark, and Germany. Transp Rev 2008, 28:495–528.

59. Department for Transport: Transport Statistics Bulletin. National Travel Survey: 2007 Interview Data. London: Department for Transport; 2008.

60. Moritz W: Survey of North American Bicycle Commuters: Design and Aggregate Results. Transp Res Rec 1997, 1578:91–101.

61. Pucher J, Dijkstra L: Promoting safe walking and cycling to improve public health: lessons from The Netherlands and Germany. Am J Public Health 2003, 93:1509–1516.

62. Danish Ministry of Transport: Danish National Travel Surveys. Copenhagen: Danish Ministry of Transport; 2007.

63. Statistics Netherlands: Transportation Statistics. Amsterdam: Statistics Netherland; 2007.

64. German Federal Ministry of Transport: German Federal Travel Survey 2002 (MiD). Berlin: German Federal Ministry of Transport; 2003.

65. Land Transport New Zealand: Workplace travel plan coordinator's guide. Wellington: Land Transport New Zealand; 2007.

66. Bauman AE, Rissel C, Garrard J, Ker I, Speidel R, Fishman E: Cycling—Getting Australia Moving: Barriers, facilitators and interventions to get more Australians physically active through cycling. Melbourne: Cycling Promotion Fund; 2008.

67. Bicycle Victoria: Transport and Liveability Statement provides $72 million for riders. Melbourne: Bicycle Victoria; 2007.

68. Rissel C: Active travel: a climate change mitigation strategy with co-benefits for health. New South Wales Public Health Bulletin 2009, 20:10–13.

69. National Cycleway Project, [http://www.tourism.govt.nz/Our-Work/National-Cycleway-Project/].

70. Land Transport New Zealand: Speed Limits New Zealand: Guidelines for setting speed limits and procedures for calculating speed limits. [http://www.ltsa.govt.nz/roads/speed-limits/speed-limits-nz.html], 2003.

71. Sport and Recreation New Zealand: Sport, recreation and physical activity participation among New Zealand adults: Key results of the 2007/08 Active NZ Survey. Wellington: SPARC; 2008.

72. Ogilvie D, Egan M, Hamilton V, Petticrew M: Promoting walking and cycling as an alternative to using cars: systematic review. Br Med J 2004, 329:763.

73. Bowles H, Rissel C, Bauman A: Mass community cycling events: Who participates and is their behaviour influenced by participation? Int J Behav Nutr Phys Act 2006, 3:39.

Adolescent-Parent Interactions and Attitudes Around Screen Time and Sugary Drink Consumption: A Qualitative Study

Libby A. Hattersley, Vanessa A. Shrewsbury, Lesley A. King, Sarah A. Howlett, Louise L. Hardy and Louise A. Baur

ABSTRACT

Background

Little is known about how adolescents and their parents interact and talk about some of the key lifestyle behaviors that are associated with overweight and obesity, such as screen time (ST) and sugary drink (SD) consumption. This qualitative study aimed to explore adolescents' and parents' perceptions, attitudes, and interactions in regards to these topics.

Methods

Using an exploratory approach, semi-structured focus groups were conducted separately with adolescents and (unrelated) parents. Participants were recruited from low and middle socio-economic areas in the Sydney metropolitan area and a regional area of New South Wales, Australia. Transcripts were analysed using thematic analysis for each of the four content areas (adolescent-ST, adolescent-SD consumption, parents' views on adolescents' ST and parents' views on adolescents' SD consumption).

Results

Nine focus groups, with a total of 63 participants, were conducted. Broad themes spanned all groups: patterns of behavior; attitudes and concerns; adolescent-parent interactions; strategies for behavior change; and awareness of ST guidelines. While parents and adolescents described similar patterns of behaviour in relation to adolescents' SD consumption and ST, there were marked differences in their attitudes to these two behaviours, which were also evident in the adolescent-parent interactions in the home that they described. Parents felt able to limit adolescents' access to SDs, but felt unable to control their adolescents' screen time.

Conclusion

This study offers unique insights regarding topics rarely explored with parents or adolescents, yet which are part of everyday family life, are known to be linked to risk of weight gain, and are potentially amenable to change.

Introduction

Overweight and weight-related behaviors developed during childhood and adolescence tend to track into adulthood, with significant long-term health implications [1-3]. Consumption of sugary drinks (SDs) and levels of sedentariness in this age group are of particular concern, given that there is probable and convincing evidence, respectively, that these behaviors are associated with increased risk of weight gain and the development of obesity [3-8]. In addition, there is evidence of a clustering of obesity-promoting behaviors, through the positive association between screen time (ST) and SD intake [9-12].

National and international guidelines consistently recommend that adolescents choose water as their main drink, keep their consumption of SDs to a minimum, and limit ST to less than 2 hours per day [13-15]. However, the evidence indicates that many adolescents fail to meet healthy eating guidelines and consume excessive volumes of SDs [5,16-18]. In a 2004 estimate in Australia 60% of boys and 40% of girls aged 12-16 years reported drinking more than 250 ml per day of soft drink [16]. Longitudinal research indicates that consumption of SDs tracks from adolescence into adulthood. [19] The majority of Australian adolescents also spend over 35 hours per week in screen time, and exceed recommendations to limit small screen recreation [20] The home environment contributes to SD and ST behavior patterns, through food availability, access to electronic media and television screens, and parental modeling [17,21,22]. SD and ST were thus selected as an appropriate focus of this study as they are specific behaviours related to weight status, potentially amenable to intervention within the home environment and particularly prevalent amongst adolescents [5].

While adolescents often experience a higher level of autonomy and independence compared with younger children [22], their parents and caregivers continue to play a key role in setting boundaries and supporting healthy lifestyle choices. However, little is known about how adolescents and their parents talk about weight and associated lifestyle behaviors, even though adolescent-parent relationships are a significant area of publication and research [23]. The nature of the interaction between parents and their adolescent dependents with regards to weight and associated behaviors is likely to impact on adolescents' perceptions and practices around these issues. Understanding adolescents' and parents' attitudes, concerns, and interactions about specific weight-related behaviours is fundamental to designing communication messages and obesity prevention interventions.

Therefore, the aim of this study was to explore adolescents' and parents' (of adolescents) perceptions, attitudes, and interactions in regards to adolescents' SD and ST within the home environment. The study had parallel research questions related to adolescents' and parents' perceptions:

• Adolescents:

  (i) Are adolescents concerned about their SD consumption and recreational ST?

  (ii) To what extent are these behaviors amenable to change? How could they be reduced?

• Parents of adolescents:

  (i) Are parents concerned about their adolescents' SD consumption and ST?

  (ii) To what extent do they see these behaviors as amenable to change? How could they be reduced?

• Both:

  (i) What types of interactions occur at home regarding SD and ST? Are these topics discussed, and if so, how are they raised?

(ii) Are the views of parents and adolescents on SD and ST consistent or divergent?

Methods

Procedure

A qualitative study design using semi-structured focus groups was selected as an appropriate method of investigation for an initial exploration of perceptions and to gauge the breadth and strength of publicly expressed attitudes [24]. A market research company (MRC) was contracted to recruit participants and conduct the focus groups, in accordance with the study team's specifications. The study was approved by the Ethics Committees of The Children's Hospital at Westmead and The University of Sydney.

Participants

Male and female adolescents in high school grades 8, 9 and 10 (i.e. 13-16 year olds) and (unrelated) parents or primary caregivers with at least one child in this age range were eligible to participate. The decision to recruit unrelated parents and adolescents was made to avoid contamination, as practical considerations meant that the researchers could not guarantee there could be simultaneous focus groups with related parents and adolescents.

Participants were recruited from residences in areas classified as low to middle socio-economic status (SES) in the Sydney metropolitan region and a regional centre of New South Wales, Australia. This study focused on adolescents and parents from low-middle socioeconomic groups because these groups tend to be at higher risk of the development of overweight [25]. SES was based on The Australian Bureau of Statistics Index of Relative Socio-economic Advantage/Disadvantage (IRSAD) [26] which is calculated for postal area codes. The IRSADs for each postcode in the Sydney Metropolitan region was ranked and divided into quintiles. People residing in postcodes in the bottom three quintiles (Q) were considered to be low (Q1), low-middle (Q2), and middle (Q3) SES and eligible for inclusion. The regional centre had an IRSAD that was equivalent to the metropolitan locations.

The MRC approached people listed in its database to participate in the study if they: had given consent to future contact; met the target group specifications; and had not participated in any form of market research in the past six months. Using their standard recruitment methods, the MRC contacted 457 parents by and a further 33 by phone. In order to enhance recruitment in the regional centre, a subsequent was sent to the remaining 187 non-eligible residents listed in the MRC's database with information about referring family and friends to the study. All participants received written information explaining the study. Informed consent by adolescents and by parents was a requirement for the study and participants were reimbursed for their travel expenses and time. Participants were required to speak fluent English in order to be able to participate fully in the focus group discussions.

Data Collection

Focus group discussion questions and prompts were developed to address the research questions, in a way that would promote interest and open discussion (see Table 1). Separate focus groups were conducted with parents and adolescents. Separate male and female adolescent groups were held as it was expected that adolescents would feel more comfortable discussing the study topic with peers of the same sex. Parent groups were arranged to comprise two combined mother-father groups and two separate mother and father groups. This arrangement was designed to ensure representation and active participation by fathers as well as mothers, recognizing that few studies actually seek out fathers.

Table 1. Discussion prompts

Parent focus groups
What do you think about the amount of time your child spends on screen-based leisure activities/their consumption of sugary drinks?
Tell me about any agreements, disagreements or rules you've had with your child about their use of screen-based leisure activities/consumption of sugary drinks?
If you wanted your child to reduce the amount of time they spend on screen-based leisure activities/the amount of sugary drinks they drink, what kind of strategies do you think would be effective?
Do you know whether there are any guidelines on the maximum amount of time adolescents should spend on screen-based leisure activities?

Adolescent focus groups
What do you think about the amount of time you spend watching TV/videos/DVDs or on computer games (not including time spent for educational purposes)?
How much do you care about your intake of sugary drinks?
Tell me about any agreements, disagreements or rules you've had with your parents about your use of screen-based leisure activities/consumption of sugary drinks?
What things would make it easier for people your age to drink less sugary drinks?

Each focus group session was conducted over 90-120 minutes and was held in a community-based meeting room. All sessions were digitally audio-recorded with participants' consent. A researcher from the study team attended and observed each session.

Data Analysis

De-identified transcripts of the audio-recordings, typed verbatim, were provided by the MRC and checked for quality by a member of the research team. The transcripts were manually analyzed using thematic analysis. Three members of the research team independently read the transcripts and discussed the key ideas and common themes arising across the groups, and spanning the four content areas: adolescent-ST; adolescent-SD; parents' views on adolescents' ST; and parents' views on adolescents' SD. Following agreement on a draft coding structure, two members of the research team independently coded the data for each of the four content areas. The five broad themes comprised:

- Patterns of behavior
- Attitudes, beliefs and concerns;
- Adolescent-parent interactions;
- Strategies for behavior change;
- Awareness and perception of ST guidelines (the study did not specifically inquire regarding SD guidelines as there are no clear, precise guidelines).

A third member of the research team checked the results within each content area for consistency and minor revisions were made to the coding structure where appropriate. Consensus on the final coded data was reached in a straightforward manner and summaries of the findings were checked by members of the research team for accuracy. The five themes remained intact and it was agreed that they accurately captured the key information in the entire dataset, across all groups and content areas of interest, and in a way that addressed the overall research questions.

Results

Response Rate

A total of 402 people responded to the initial study invitation, of whom 103 met the eligibility criteria. A final sample of 63 participated in a focus group (63% participation rate; 31 adolescents and 32 parents). A total of nine focus groups (five with adolescents; four with parents) were held in March 2008. Following completion of the nine focus groups, the primary and co-facilitator agreed that response saturation had been attained.

Participant Characteristics

Groups were primarily organized according to geographic location, with six in metropolitan Sydney and three in the regional centre (see Table 2). Adolescent participants in each group ranged in age between 13-16 years (Australian school Grades 8-10); 42% were female. The mean age of parent participants was 46 years for mothers and 43 years for fathers, and 63% were mothers. Approximately 40% of parents had finished high school or less as their highest education level, half had attained a tertiary certificate or diploma and just over 10% held a tertiary degree. Almost 20% of parent and 13% of adolescent participants were born outside Australia; and about 40% of adolescents were from families where at least one parent was born overseas. Almost one-third of adolescents and parents of adolescents lived in areas classified as low or low to middle SES, with the remainder of participants residing in areas ranked as middle SES.

Table 2. Focus group configuration

| Group | Location | Adolescents | | Parents | |
		Male (n)	Female (n)	Fathers (n)	Mothers (n)
1	Metropolitan		8		
2	Metropolitan				8
3	Regional centre		5		
4	Regional centre	5			
5	Regional centre			2	6
6	Metropolitan	8			
7	Metropolitan			8	
8	Metropolitan	5			
9	Metropolitan			2	6
TOTAL		18	13	12	20

Focus Group Discussion Themes

The key findings and supporting quotations in relation to the major themes for each of the four content areas of interest were tabulated as a way of highlighting commonalities and differences between parents and adolescents, and between the behaviors of interest [see Additional file 1]. The range of comments in relation to these themes, across adolescents and parents and ST and SDs, are summarized below.

Behavior Patterns

While carbonated soft drinks were largely discussed by both parents and adolescents as a 'treat' reserved for weekends, visitors or special occasions, other sugary drinks such as fruit juice and milk-based drinks were discussed as a daily component of adolescents' diets.

ST was overwhelmingly discussed as a part of daily life for adolescents. Parents also saw it as part of adolescents' daily life for communication and homework, and provided vivid and detailed descriptions of the times, occasions, purposes and preferences of their adolescents for using ST.

Attitudes, Beliefs and Concerns

For both ST and SD consumption, parents expressed noticeably more health-related concerns than adolescents.

Parents' concerns relating to SDs centered on the sugar and caffeine content, as well as acidity of various drinks, and resultant impacts on dental and general health. Interestingly, however, these concerns were not intensely or personally expressed, and not related to weight. Adolescents, on the other hand, expressed some mild health concerns relating to sugary drinks, including their contribution to weight.

Parents expressed a high degree of concern and frustration regarding their adolescent's ST. This was the foremost response to the topic in the parent groups. Parents were particularly concerned with the impact of excessive ST on social interaction and social skills, as well as competing with homework and learning. A few parents also raised concerns relating to ST and radiation exposure, repetitive strain injury and corruption of spelling. However, many of the parents accepted their adolescents' time spent in screen activities as inevitable and described themselves as 'giving up' and accepting ST as a feature of family life and society generally.

A common response from both parents and adolescents was that adolescents cannot be relied on to self-moderate their intake of SDs or use of ST. Further, parents frequently expressed a belief that they had little, or no, control over their adolescents' behaviors relating to ST. A common perception was that adolescents would ignore them, or go 'behind their back.' This was supported by comments from several adolescents who suggested that they would ignore or overcome any rules which were set with relative ease.

Adolescent-Parent Interactions

The nature of interactions between adolescents and parents varied depending on the behavior in question. Interactions regarding SDs were described as discussions where parents reminded the adolescent of what they should be drinking. In contrast, conversations focusing on ST were commonly described as being confrontational, with parents attempting to enforce rules. Both parents and adolescents noted that these discussions around ST frequently escalated into arguments and shouting.

Role modeling was identified as a relevant issue by some adolescents and parents in relation to SDs.

Strategies for Behavior Change

Common strategies identified by parents for promoting healthy behaviors in their adolescent children, in terms of SDs and ST, included controlling availability or access, and providing alternatives (particularly water). The adolescents perceived that this could work for SDs, but restrictions and enforced rules would be the best strategy for changing ST.

ST Guidelines

There was limited awareness among both parents and adolescents about the guideline to limit small screen recreation time to less than 2 hours/day [13]. The question on this topic generated lots of different ideas about the rationale for a guideline. Mostly, parents thought that the 2 hour/day guideline wasn't feasible, was 'out of touch' and certainly would not be acceptable to adolescents. Some parents queried and derided the source or basis for the guideline and, in one group, there was an explicit dismissal of health guidelines generally. Some parents were concerned about how the guideline could be achieved in practice. Adolescents responded as parents anticipated, by rejecting the idea of a 2 hour/day limit as unrealistic and unacceptable.

Participants' Discussion Style

While the major themes refer to discussion content, the analysis also identified differences between parents and adolescents in their style of discussion. Overall, parent focus groups involved more lively discussion than did the adolescent groups. Adolescents were slower to open up in discussion, and in some cases tended to express what might be taken as a publicly expected view, such as bragging about their ability to circumvent parental rules. In many ways, the issue of control was a present and overarching theme, with parents expressing the extent to which they can and cannot control their adolescent's behavior, and adolescents commenting on the extent they do and do not go along with parental control.

Discussion

While the discussions on ST and SD consumption addressed similar themes (i.e. in terms of attitudes, interactions and behavior patterns), the nature of the attitudes and adolescent-parent interactions were quite different for each of these behaviors. Overall, neither parents nor adolescents expressed strong personal concern about adolescents' SD consumption, yet adolescents' ST elicited considerable concern amongst parents and was associated with family conflicts. Thus, these different behaviors had quite different emotional salience to parents and adolescents. Research with parents and adolescents has described a complex mix of factors, such as social expectations and marketing, parenting style and ability to say 'no,' and communication taboos, that mediate perceptions and communications about weight [27]; some of which may be at play in this study.

Findings from this study that parents feel able to limit their children's access to SDs appear encouraging, given that previous research indicates that the home environment is critical in shaping children's dietary and activity behaviors [17,28,29], and that adolescents themselves recognize the importance of the home environment in influencing their behaviors [30]. However, there may be some discrepancy between parents' low level of personal concern and adolescents' actual consumption [16]. This mismatch, where 'treats' may in fact be consumed frequently, poses a barrier for change and needs to be anticipated in any communications or family interventions [27].

On the other hand, the extent to which parents felt unable to control their children's ST was concerning. For example, parents described how they had given up trying to limit ST, as a result of failed or negative interactions. Nevertheless, adolescents indicated that enforced rules were an appropriate control strategy. Comparing the adolescent and parent responses suggests that parental rules about ST are likely to lead to conflict, unless perhaps if rules are established when children are younger. No-one mentioned the use of TV time monitors or similar devices as a method for monitoring or enforcing rules.

While role modeling was mentioned in relation to SDs in both parent and adolescent groups, the limited emotional engagement with the topic suggests that its influence on adolescents continues to be under-estimated, especially by parents [27]. Research has highlighted that parents can be positive role models with their adolescent children in regard to weight-related behaviors [17,29,31]. There appears to be scope to promote the significance of role modeling and its applicability to influencing both SD consumption and ST to parents, through parent-targeted health communications and skill-based parenting interventions.

The low level of awareness of health recommendations and guidelines on recreational ST, and the dismissive response to the idea of guidelines, suggests that this guideline challenges public views and actual practices, which is consistent with findings from US qualitative research [32] and the context where Australian and North American family households have high access to electronic media. Recent data shows that 99% of Australian households with young people (8-17 years) have at least one television set, 98% have at least one computer or laptop, 97% a DVD player and 91% have an internet connection [21]. Clearly the health reasons for ST recreation guidelines, such as the association between recreational ST and fitness, is an important underpinning that should be featured in public health interventions and communications [33].

The typical roles that adolescents and parents play in relation to each other, with disagreements over everyday issues and struggles around control [23], were apparent in this study, even though the parents and adolescents were not related to each other and the group discussions were held separately. In this study parents frequently expressed issues in relation to their ability to control adolescents' behavior, and adolescents frequently asserted their independence from parental controls. While both positions correspond to an accepted public image, they also provide a reminder that any health communications and skill-based parenting interventions related to adolescents must always recognize and respond to this fundamental dynamic.

Overall, the study shows that there is scope to influence adolescents through their home environment and parental use of positive strategies, such as setting and applying defensible rules, limiting access and availability and role modeling. While an exploratory study, the findings suggest directions for health communication messages to parents and the value of structured parenting interventions, which typically focus on parents' understanding and acceptance of adolescent development as well as communication and listening skills [34].

While the study was generally successful in recruiting people from lower socioeconomic areas, there was difficulty in recruiting and retaining some participants from the most disadvantaged areas. The use of market research company recruitment methods may have limited contacts with people from low socio-economic backgrounds, although most eligible households are known to have internet access [21] and the purposive targeting of low SES areas meant that any potential bias was redressed. The study participants came from backgrounds that were fairly representative of the Australian population; although like many research studies, the views of the most disadvantaged, Aboriginal people or a wide range of people from culturally and linguistically diverse cultures are not comprehensively represented. Further research targeting these groups is required to investigate their particular environmental, cultural and material circumstances. While there is some possibility that participants were more health conscious than their peers, the recruitment information referred to 'lifestyle' rather than 'health' and health did not appear as a dominating theme. Overall, this study was successful in recruiting adolescents and parents from low and middle SES areas, with a particular strength in the participation of fathers, given that they are often under-represented.

The focus group design itself may have incurred some limitations, especially in relation to adolescents, who displayed a limited degree of 'conversational competency' in some cases [35] and tended to produce socially expected responses. However, it is important to understand and respond to these public views in designing communication and other interventions. Deeper understanding on specific issues, such as discrepancies between parents' level of concern and adolescents' actual behaviors, in relation to both SD and ST, would require additional research using a mix of methods, with related parent and adolescent participants.

Nevertheless, the study provides an initial exploration and offers insights regarding topics rarely explored in relation to adolescents, yet which frequently occur within the family and home environment, are known to be linked to risk of weight gain, and are discrete behaviors that are potentially amenable to change. The findings complement previous qualitative research with parents of younger children [27] and adolescents [30], as well as quantitative studies on these topics (for example [10,12,17,20]), thus providing guidance on angles and perspectives that can be used by policy makers, health practitioners and researchers to guide interventions and other health communications.

Competing Interests

The authors declare that they have no competing interests.

Authors' Contributions

Libby Hattersley (LHatt), Lesley King (LK), Vanessa Shrewsbury (VS), Louise Hardy (LH) and Louise Baur (LB) all contributed to the development of research questions and the design of the study, and the Ethics Committee submission. VS, LK and LHatt were involved in contracting the market research company and providing specifications and protocols for recruitment and conduct of groups. VS attended all focus groups and checked all transcripts. VS, LHatt, LK and Sarah Howlett (SH) undertook data coding and initial analyses; LH, LB, VS and SH were involved in checking the coding and analysis. All authors contributed to the interpretation of data and the writing of the manuscript. All authors read and approved the final manuscript.

Acknowledgements

The authors thank all of the parents and adolescents who participated in the study and the Ipsos-Eureka Social Research Institute who were contracted to conduct the fieldwork. The study was funded by the NSW Department of Health.

References

1. Lobstein T, Baur L, Uauy R: Obesity in children and young people: a crisis in public health. Obes Rev 2004, 5(Suppl 1):4–104.

2. Wang LY, Chyen D, Lee S, Lowry R: The association between body mass index in adolescence and obesity in adulthood. J Adolesc Health 2008, 42(5):512–8.

3. World Health Organization: Diet, nutrition, and the prevention of chronic diseases:Report of a Joint WHO/FAO Expert Consultation. In 916 ed. Geneva: World Health Organization; 2003.

4. Crespo CJ, Smit E, Troiano RP, Bartlett SJ, Macera CA, Andersen RE: Television watching, energy intake, and obesity in US children: results from the third National Health and Nutrition Examination Survey, 1988–1994. Arch Pediatr Adolesc Med 2001, 155(3):360–5.

5. Gill TP, Rangan AM, Webb KL: The weight of evidence suggests that soft drinks are a major issue in childhood and adolescent obesity. Med J Aust 2006, 184(6):263–4.

6. Ludwig DS, Peterson KE, Gortmaker SL: Relation between consumption of sugar-sweetened drinks and childhood obesity: a prospective, observational analysis. Lancet 2001, 357(9255):505–8.

7. Tam CS, Garnett SP, Cowell CT, Campbell K, Cabrera G, Baur LA: Soft drink consumption and excess weight gain in Australian school students: results from the Nepean study. Int J Obes (Lond) 2006, 30(7):1091–3.

8. Vartanian LR, Schwartz MB, Brownell KD: Effects of soft drink consumption on nutrition and health: a systematic review and meta-analysis. Am J Public Health 2007, 97(4):667–75.

9. Kremers S, Horst K, Brug J: Adolescent screen-viewing behaviour is associated with consumption of sugar-sweetened beverages: The role of habit strength and perceived parental norms. Appetite 2007, 48(3):345–50.

10. Salmon J, Campbell KJ, Crawford DA: Television viewing habits associated with obesity risk factors: a survey of Melbourne schoolchildren. Med J Aust 2006, 184(2):64–7.

11. Van den BJ, Van MJ: Energy intake associated with television viewing in adolescents, a cross sectional study. Appetite 2004, 43(2):181–4.

12. Vereecken CA, Todd J, Roberts C, Mulvihill C, Maes L: Television viewing behaviour and associations with food habits in different countries. Public Health Nutrition 2006, 9(02):244–50.

13. The Australian College of Paediatrics: Policy statement. Children's television. J Paediatr Child Health 1994, 30(1):6–8.

14. American Academy of Pediatrics: Committee on Public Education. Children, Adolescents, and Television. Pediatrics 2001, 107(2):423–6.

15. Smith A, Kellet E, Schmerlaib Y: Australian Guide to Healthy Eating. Commonwealth of Australia. Canberra: AGPS; 1998.

16. Booth ML, Okely AD, Denney-Wilson E, Hardy LL, Yang B, Dobbins T: NSW Schools Physical Activity and Nutrition Survey (SPANS) 2004: Full Report. Sydney: NSW Department of Health; 2006.

17. Campbell KJ, Crawford DA, Salmon J, Carver A, Garnett SP, Baur LA: Associations between the home food environment and obesity-promoting eating behaviors in adolescence. Obesity (Silver Spring) 2007, 15(3):719–30.

18. Hector D, Rangan A, Louie J, Flood V, Gill TP: Soft drinks, weight status and health: a review. Sydney: NSW Centre for Public Health Nutrition (now know as Cluster of Public health Nutrition, Prevention research Collaboration, University of Sydney); 2009.

19. Lien N, Lytle LA, Klepp KI: Stability in consumption of fruit, vegetables, and sugary foods in a cohort from age 14 to age 21. Prev Med 2001, 33(3):217–26.

20. Hardy LL, Dobbins T, Booth ML, Denney-Wilson E, Okely AD: Sedentary behaviours among Australian adolescents. Aust N Z J Public Health 2006, 30(6):534–40.

21. ACMA: Access to the internet, broadband and mobile phones in family households. Australian Communications and Media Authority; 2008.

22. Story M, Neumark-Sztainer D, French S: Individual and environmental influences on adolescent eating behaviors. J Am Diet Assoc 2002, 102(3 Suppl):S40–S51.

23. Smetana JG, Campione-Barr N, Metzger A: Adolescent development in interpersonal and societal contexts. Annual Review of Psychology 2006, 57(1):255–84.

24. Willis K, Green J, Daly J, Williamson L, Bandyopadhyay M: Perils and possibilities: achieving best evidence from focus groups in public health research. Aust N Z J Public Health 2009, 33(2):131–6.

25. Shrewsbury V, Wardle J: Socioeconomic status and adiposity in childhood: a systematic review of cross-sectional studies 1990-2005. Obesity (Silver Spring) 2008, 16(2):275–84.

26. Australian Bureau of Statistics: Census of Population and Housing: Socio-economic indexes for areas (SEIFA) Australia 2006. Canberra, Australian Bureau of Statistics. Information paper(Australian Bureau of Statistics); 2006. Ref Type: Internet Communication.

27. Pagnini D, King L, Booth S, Wilkenfeld R, Booth M: The weight of opinion on childhood obesity: recognizing complexity and supporting collaborative action. Int J Pediatr Obes 2009, 1–9.

28. Grimm GC, Harnack L, Story M: Factors associated with soft drink consumption in school-aged children. J Am Diet Assoc 2004, 104(8):1244–9.

29. Hardy LL, Baur LA, Garnett SP, Crawford D, Campbell KJ, Shrewsbury VA, et al.: Family and home correlates of television viewing in 12–13 year old adolescents: the Nepean Study. Int J Behav Nutr Phys Act 2006, 3:24.

30. Booth ML, Wilkenfeld RL, Pagnini DL, Booth SL, King LA: Perceptions of adolescents on overweight and obesity: the weight of opinion study. J Paediatr Child Health 2008, 44(5):248–52.

31. Horst K, Oenema A, Ferreira I, Wendel-Vos W, Giskes K, van Lenthe F, et al.: A systematic review of environmental correlates of obesity-related dietary behaviors in youth. Health Educ Res 2007, 22(2):203–26.

32. Jordan AB, Hersey JC, McDivitt JA, Heitzler CD: Reducing children's television-viewing time: a qualitative study of parents and their children. Pediatrics 2006, 118(5):e1303–e1310.

33. Hardy LL, Dobbins TA, Denney-Wilson E, Okely AD, Booth ML: Sedentariness, small-screen recreation, and fitness in youth. Am J Prev Med 2009, 36(2):120–5.

34. Henricson C, Roker D: Support for the parents of adolescents: a review. J Adolesc 2000, 23(6):763–83.

35. Warr DJ: "It was fun... but we don't usually talk about these things": Analyzing Sociable Interaction in Focus Groups. Qualitative Inquiry 2005, 11(2):200–25.

Women with Postpartum Depression: "My Husband" Stories

Phyllis Montgomery, Pat Bailey, Sheri Johnson Purdon, Susan J. Snelling and Carol Kauppi

ABSTRACT

Background

The research on Postpartum Depression (PPD) to date suggests that there is a knowledge gap regarding women's perception of their partners' role as carer and care activities they perform. Therefore, the purpose of this study was to describe women's understanding of their partners' or husbands' involvement in the midst of PPD.

Methods

This study used interview data from a larger study of northern and rural Ontario women's stories of help-seeking for PPD. The interpretive description approach was used to illustrate the complexity of women's spousal connections in PPD. Data from a purposive community sample of 27 women who self-identified as having been diagnosed with PPD was used. From the verbatim *transcribed interviews a number of data excerpts were identified and labeled as "my husband" stories. Narrative analysis was employed to examine these stories.*

Results

During this time of vulnerability, the husbands' physical, emotional and cognitive availability positively contributed to the women's functioning and self-appraisals as wife and mother. Their representations of their husbands' 'doing for' and/or 'being with' promoted their well-being and ultimately protected the family.

Conclusion

Given that husbands are perceived to be central in mitigating women's suffering with PPD, the consistent implementation of a triad orientation, that includes woman, child and partner rather than a more traditional and convenient dyadic orientation, is warranted in comprehensive postpartum care. Finally, this study contributes a theoretical understanding of responsive as well as reactive connections between women and family members during the postpartum period.

Background

The complex relationship between Postpartum Depression (PPD) and the marital relationship is well-documented in the nursing literature. Numerous studies have consistently associated marital factors such as conflict, dissatisfaction and support with the risk for PPD [1]. Beck [2] in a meta-analysis of 84 studies published during the 1990s, found that marital satisfaction had a moderate predictive relationship with PPD; this finding is supported by O'Hara and Swain's earlier meta-analysis [3]. Such findings emphasize that, for couples with strained relationships, the transition process associated with childbirth is that much more challenging.

During this transition, family integrity may be compromised by the presence of PPD. Maternal depression has been commonly associated with negative relational distress such as fear of being rejected by the partner, misdirected anger, withdrawal, martial dissatisfaction as well as poorer communication of needs and expectations [4,5]. Paley et al. [6] contend that during this period of adjustment it is typical for the developing partners and parent-infant relationships to become stressed. For successful relational adjustments the presence and involvement of both partners was found to be essential. Dressel and Clark [7] reported that PPD also threatens interdependence within couples by challenging their notions of family care dynamics. Further, according to Beck [8] the traditional family care role expectations required readjustments related to men's role as 'carer' for their partner.

Some authors have suggested, however, that this carer role has been overlooked by health professions; an omission partly attributed to postpartum nursing's traditional focus on the mother and infant dyad rather than on a family perspective

[9-11]. Others have found that care giving is also seldom associated with the traditional ideals of masculinity [12,13]. These authors determined that women with PPD nevertheless sought partners' or relatives' assistance to lessen their suffering and to assist them with functioning. Page and Wilhelm [14] found that a woman's perception of spousal support in depression was positively related to a decrease in reported symptomatology. Their conceptualization of relationship quality, however, was imprecise.

In summary, the research to date suggests that there is a knowledge gap regarding women's perception of their partners' role as carer and their involvement in care activities [1,5,11,15]. Although there are several quantitative studies examining relationships among maternal depression, marital conflict, support and/or satisfaction [1,16], less is known about how women perceive their spousal interactions in the context of a mental disorder such as PPD [5,13,15,17]. Bost et al. [16] have reported that women's perceptions of relational experiences with husbands tend to be among the most salient predictors of their adjustment and well-being. Therefore, the purpose of this study was to describe women's understanding of their partners' or husbands' involvement in the midst of PPD. This study used interview data from a larger study of northern and rural Ontario women's stories of help-seeking for PPD.

Literature Review

PPD is a serious, non-psychotic condition affecting approximately 13% of new mothers globally [18]. Oats et al. [19], in a descriptive qualitative study, explored PPD attitudes and beliefs of new mothers, their husbands, relatives and health care providers in 15 health centers located in 11 countries. Across all centers, participants described 'morbid unhappiness,' a condition comparable to Western's Diagnostic and Statistical Manual of Mental Disorders, 4th edition, Text Revision (DSM-IV-TR) diagnosis of PPD [20]. In eight of the 11 countries, participants viewed a combination of stressed marital and family relationships, women's fatigue, and perceived lack of spousal support as being associated with the experience of maternal depression. These researchers recommended the importance of interventions aimed at strengthening the immediate family's acceptance and understanding of this universal condition. Consistent with this view, Rodrigues, Patel, Jaswal and deSouza [21] found that non-Western and Western women describe their PPD experiences through a social rather than a biomedical lens. That is, women, regardless of culture, usually interpret their postpartum emotional distress in the context of the perceived quality of their social connections. Scrandis [15] found that women with PPD described these positive connections in terms of mutual empathy and empowerment; these were dynamics that allowed them to interpret their symptoms and to cope.

There is a preponderance of evidence suggesting that positive marital relationships protect women's health in the postpartum period [14,22,23]. For example, Dennis and Ross [1] examined the relationship between 396 women's perceptions of husband support and the development of depressive symptoms. The measure to assess the dimensions of support, inclusive of appraisal/emotional, informational and instrumental, was based on literature and expert review exclusive of women with PPD. These authors suggested that women were more likely to develop depressive symptoms if they perceived that their husbands socially excluded them, discouraged them from seeking help, or did not recognize their efforts to nurture the infant at two months postpartum [1].

Further, in their examination of paternal support and maternal depression, Smith and Howard [24] claimed that the types and amount of paternal support change during a couple's first two years as new parents. They found that, at four months, nearly 57% of over 582 women reported that their husbands provided at least five types of support as compared to 45% at 24 months. Change in support was related to maternal depressive symptomatology. At 24 months, decreases in support were related to increases in depression. Decreases in support were unexpectedly associated with decreases in depression at four months. These researchers suggested that the lessening of protective processes provided by men during this early transition do not function as anticipated. This confounding result may partly be explained by Bell et al.'s [25] work with 18 new parents. They suggested that couples were often involved in particularly complex, "messy," nonlinear relational processes in the early postpartum period, a required phase of developing new ways of being together as partners and parents.

A number of researchers have attempted to understand mothers' complex subjective experiences in PPD [5,26,27]. Beck [8] conducted a meta-synthesis of 18 qualitative studies based on 309 women's experiences of PPD. She suggested that across this work the experience described by the women followed an individual trajectory influenced by biopsychological as well as social processes. Many of the included studies (61%) appeared in the nursing literature. Across the studies four themes identified by 309 women were incongruity between expectations and realities of motherhood, spiraling downward, pervasive loss and making gains. With respect to their role as wife, women were ashamed to approach their partners for help. If they disclosed their symptoms to their partners, women often characterized their partners' response as non-comforting. This, in turn, contributed to their sense of isolation and strained marital relations.

In subsequent work, Everingham, Heading and Connor [28] explored six couples' understanding of their experience of PPD. They found that all women identified a priority need to have a partner's understanding of their emotional distress. In turn, they contended that understanding lessened relational conflict as well as protected women against being labeled as 'incompetent' mothers. Although husbands recognized their spouses' need for understanding, they found that husbands often misunderstood their partners' intended messages. Expressions of distress were interpreted through a physical, personality, or psychological lens rather than through the wives' frame of being the 'good' mother. Couples' understanding of PPD from divergent frames of reference appeared to impede their adjustment during this inherently stressful transition period. Tammentie et al. [5] studied the association between women's depressed mood and family dynamics in 389 two-parent families. They found that mothers with depressive symptoms generally reported less positive family dynamics compared to their partners. Both women and their partners reported less mutuality compared to non-depressed couples. These research studies build on the work of Bost, Cox, Burchinal and Payne [16] who asserted that in the presence of PPD, husbands' and wives' perceptions of relational "give-and-take" are incongruent.

A number of researchers suggest that the emotive and social implications for husbands whose spouses are diagnosed with PPD include depression, helplessness, isolation, loss of intimacy, increased marital strain and worry [5,29-31]. In another study of men whose partners have been diagnosed with PPD, men described loss of both their spouse and their once known intimate reciprocal relationship [32]. Familiar patterns of spousal interacting and addressing problems became problematic and despairing in the context of PPD. Men's efforts 'to fix' their family situation were either not acknowledged or negatively appraised by their partner [10,33]

Methods

Design

The design of this study is a qualitative secondary analysis of data collected in a larger study focusing on women's help-seeking experiences. The original narrative study was a collaborative project involving women from a PPD peer support group, a public health agency and a school of nursing. The purpose was to describe help-seeking for PPD from the standpoint of the women themselves. This study's secondary analysis was a retrospective interpretation [34]; an approach involving engagement with the data beyond the self-evident—including both the assumed knowledge and what has already been established—to see what else might be there [35]. This was particularly relevant since data about "my husband" had captured the research team's attention in the primary study, but had not been fully examined. The what else in the "my husband" data set were women's perceptions of their connections with their partners relative to help-seeking. To interpret this inter-relational dynamic in women's efforts to seek help for PPD, questions such as: "What is happening here?" and "What might this mean?" [35,36] guided the researchers through this secondary.

Setting

The study was conducted in a northeastern Ontario region with approximately 158,000 residents, 10% which are women between the ages of 20 - 34 years of age ([37]http://www.statcan.gc.ca/). This mid-sized urban community includes the total population of smaller rural regions that range from 1,000 - 9,000 residents. A priority regional health care need is the development of a comprehensive women's wellness program ensuring a variety of services including access to appropriate mental health support and counseling [38].

Sample

In the primary study, purposive and snowball sampling techniques were used to recruit a community-based group of women. Verbal and written announcements about the study were distributed to a postpartum peer support group, a community mental health agency, a public health agency and family play centers. The inclusion criteria for the women were: 18 years of age or older, French or English speaking, self-identified as having received a diagnosis of PPD and had sought services for this condition. The sample for this analysis included 27 women in their early 20's to mid 30's, all of whom had sought medical services for their PPD symptoms within 12 months of their children's birth. As a result, all of the women were prescribed medication. As a family, all of the women were living with a partner and parented one to four children. The length of their spousal relationships varied from one to ten years. Sixteen (59%) of the participants identified themselves as living in communities with a population of less than 9,000 residents. All of the women spoke of being unprepared for the pervasiveness of PPD in relation to the physiological, psychological, emotional, social and existential aspects of their lives. Women described the onset of their symptoms in relation to a specific time period and their perception of their husbands' response (see Table 1, 2 and 3). Three women indicated that they had symptoms of depression prior to becoming pregnant, seven women reported the onset of symptoms during pregnancy and 17 (63%) described the appearance of symptoms following the baby's birth.

Table 1. Mothers reporting pre-pregnancy onset of depressive symptomatology and initial reference to husband

Mother	Onset of Symptoms	Initial Reference to Husband
M2	...within the first 12 hours I knew that I was not doing well/like it was just that quick	I was pretty ill so my husband could probably tell a lot more about what happened ... going home [from hospital] I expected it would get better
M15	I struggled with [undiagnosed] depression from childhood/[my child] was a planned pregnancy ... and then afterwards [after birth] there was a sudden change to where I'm speechless	I remember when we [her and partner] first got there [at emergency department] there were no beds ... so anyway we just - we're just gonna go home/[a family support person] came and sort of smoothed things over
M26	... after my [first child] was born/I saw [name of a previous psychiatric] by the time she was six months old/this depression and anxiety panic comes a lot at night came I wake up and just think/ "what am I going to do?"	just before [child's] first birthday my husband lost his job

Table 2. Mothers reporting prenatal onset of depressive symptomatology and initial reference to husband

Mother	Onset of Symptoms	Initial Reference to Husband
M5	... I believe that I was depressed during the pregnancy but no one identified it/six months after [my baby] was born was the critical time ...	... it was only my husband that knew what was going on. ... like it was pretty stressful kind of time for us/trying to get through it
M7	...the whole pregnancy I felt like I didn't have any control ... after coming home [from hospital] like so depressed ... I didn't feel right	I did hold them [thoughts, feelings] in and then I told my husband because he said "you know/you're not/you know you're changing ... I don't know how to explain... Talk to the doctor"
M9	When I was pregnant with [my child] um I was very detached/early not attached ... it started from me when [my child] was born	... I just keep telling [name of partner] I am not ready to have a baby [following two family losses] and then unplanned we got pregnant
M11	Even before I had the baby I look back in my journal ... a month before/a month before I had the baby	[My husband] kind of backed off like I think he didn't know what was going on/he knew what was happening all along but he didn't say anything
M12	... about the second trimester of my [first] pregnancy/crying all the time um had suicidal thoughts ... it took me about nine months to finally seek help from the doctor	I was still crying and tearful and moody ... like we [my husband and I] kind of noticing the signs now/obviously
M18	My seventh month of being pregnant I noticed that I wasn't excited/I was really nervous and fearing everything .../so anyway/I had a good delivery and even the same day/the same evening I didn't want to see [my child] ... when I came home [from hospital] I'm thinking my world just fell apart	... my husband stayed with me while I was in the hospital/also I kinda let him help out/we do almost everything ... [when child was about one month of age] I finally broke down and told my husband there was something wrong because he was went back to work and I couldn't be alone because I feared everything
M22	I first became aware of depression when I was six months pregnant	I remember looking at my baby who was crying and my husband was trying to comfort her/I had my first panic attack ... my husband went for help

Table 3. Mothers reporting postpartum onset of depressive symptomatology and initial reference to husband

Mother	Onset of Symptoms	Initial Reference to Husband
M1	... around 4 1/2 months and ah I started to recognize that there was something terribly wrong	I told [my husband] I hadn't slept in two weeks ... and my husband ... he drove me to the hospital
M3	... the first month ...the mood ... sit there by myself and cry and then feel guilty ... I suffered from a lot of anxiety	... my husband he was pretty much caring for [the baby] through the night/I would care for [the baby] during the day and as soon as he would come home I was out of the picture
M4	... I knew I needed help at three month and I waited for the four month checkup with the baby	... make everything look perfect on the outside ... [the doctor] asked me how much I was sleeping ... my husband said "What?"/he had no clue that I wasn't sleeping
M6	I was getting the anxiety that you know ... but I know one night I was really bad like ... close to four months [postpartum]	Then one night I thought I heard her cry all night/my husband said she didn't cry at all and so I guess my nerves were really bad and then one day I said like to my husband/"I couldn't/I just couldn't deal"
M8	... [my child] was 13 months and um started having real anxiety about being alone with [my child] and um feeling very overwhelmed with her care	... my husband was out of town overnight so I knew he wasn't going to be at the house that night and I was just going to be on my own and I was just very overwhelmed
M10	It started pretty much when [my child] was born and it didn't take long for me to know that it was bad	I tried to tell my husband but just like he kinda tried to deny it
M13	I had depression right of the bat like within the first month/I didn't realize when it was happening/I just knew that I was stressed out and I felt different	My husband involved my mother
M14	2 1/2 days after I had the baby/that evening I had a panic attack/I didn't know what it was	... next day [my baby] had an appointment with the doctor/my husband and I explained what happened ... that night it happened again but much worse/my husband had to call someone over
M16	I had a bit of postpartum with my first ... with my second and in fact sought help from a doctor then/I knew right away what it was with the third one	My husband could see it in me and you know he would ask me you know/"what's going on with you and the more supportive he got the more I wanted to cry"
M17	When the baby was eight weeks/exactly eight weeks/I was having a lot of physical symptoms	... so the realization [concerning PPD] came just mainly by myself and just talking to my husband
M19	When the baby was born I just didn't feel right/I didn't feel myself	... my husband had to take me to the doctor's because I knew something wasn't right
M20	... so much pressure seemed to be on me [after my second]/I got to a breaking point where I just I can't do this anymore	I know my husband before me came to the conclusion/was telling me/"there's something wrong/there's something wrong"/but I was like/"no"
M21	...during the first year [of child's life]/I didn't know it was depression .../I thought I was crazy/I thought there's something wrong with me mentally	I didn't show it [depression] in from of him .../he thought it was just the baby blues ... he/he did notice that I was/there was something wrong with me ... like he didn't have anything like that in his family
M23	... I was so sick right after she was born. I was so depressed	... it came to a point that I said to my husband/"I have to be serious/I told him/"I really get depressed and I have fear with it"
M24	Throughout the pregnancy I had never felt better in my whole life/one week after the baby was born I was sinking/I was terrified to be left alone with [the baby]	I told my husband/"don't leave me alone/I'm afraid/"his thinking was/"you are just afraid/you got to get use to [the baby]"
M25	... first two weeks of the birth of each of my two children I had moods/couldn't sleep/tired no matter what and bad nerves/I was hospitalized the second time	... after I came home from hospital [psychiatric hospitalization] I was so so anxious my husband had to take care of me and the kids
M27	... for one year after [first child's name] was born I felt like I was going crazy ... I totally changed ... I didn't feel like I was me anymore	... my my husband thought it was post/we both thought it was postpartum depression/thought it was emotion

Data Collection

Following ethical approval from two respective boards, Laurentian University as well as Sudbury & District Health Unit, and under conditions of informed consent, audio taped interviews were conducted by a mental health nurse (PM) and a PPD peer support worker (SJP). In the primary study, an unstructured interview characterized as a guided conversation was followed. This type of non-threatening, supportive exchange allowed women to share help-seeking information grounded by contextual details that made sense to them [39-41]. The topic of questions used to begin or guide the conversation were about their help-seeking decisions, their understanding of their support needs and their appraisal of available support resources. Each woman completed at least one audio taped interview, ranging in length from 30 to 90 minutes. The interviews were conducted in homes, community clinics or coffee shops. Four women requested a second interview to permit them more time to tell their stories. Since one of the interviewers may not have been a stranger to prospective participants, specific demographic data was not collected respecting participants' right to confidentiality. Women, however, shared information such as age, mental health histories, onset of PPD, and/or their and their partners' employment in formulating their help-seeking stories.

Narrative Analysis

From the verbatim transcribed interviews a number of data excerpts were identified and labeled as "my husband" stories. The label of "my husband" was the most common term of reference used by women when they spoke about their partner. The broad underlying premise of narrative inquiry within social science research is the belief that individuals most effectively make sense of their world and communicate these meanings by (re)constructing stories or narrating them [40,42-48]. Therefore, it is important to examine these structures within interview data. An eclectic narrative analysis strategy incorporating Labov and Waletzky's [49] functional analysis model of individual stories and Agar and Hobbs' [50] strategy of identifying story coherence across interviews developed by the second author (PB) was used to interpret the identified stories [51-54]. This analytical approach was used because it provides an explicit analysis process and facilitates the interpretation of meaning in complex stories [55].

In the primary study, "husband" stories were identified as first-person event-specific and generic stories but not systematically explored by the research team. For this study, using unmarked transcripts, the first two authors individually deconstructed each "husband" story into the following story elements: abstract, orientation, complicating action, resolution, evaluation and coda, as defined by Labov and Waletzky [49]. These authors then jointly reviewed and compared these elements within and across the stories to assess shared content and structure. In addition, they examined the stories for the inclusion of discursive and rhetorical devices [56]. For example, participants frequently repeated words or phrases and used direct quotations to enhance the credibility of their accounts. Finally, these authors explicated an interpretation of the story meanings (content themes) and function (utility) of these meanings to develop an understanding of husband behavior in the midst of PPD for the participants.

To demonstrate meaning and function of the husbands within the context of PPD, as described by these participants, the results are presented in two sections: the stories and a case example. Within the stories section, story structure and content as well as story meanings and function are described. Next, a case example is presented to illustrate one participant's understanding of her husband's involvement in the midst of overwhelming symptoms. The case example, presented using Labov and Waletzky's [49,57] structural elements, is intended to make the analysis process transparent [48,58]. These presentation strategies facilitate the reader's ability to examine the analytic logic and researchers' interpretation process [35,59].

Results

Each of the 27 participants shared information about their "husbands," his role as an adult partner rather than that of father during PPD. In total, the women contributed 153 stories about their partners' availability. Using explanatory narratives [55] to communicate complex points of view that might be otherwise hidden to the outsider, the women characterized two types of husband availability: 'doing for' and 'being with.' These accounts illustrate their insider perceptions about their husbands' presence, both tangible and intangible. The data in each section is presented in an un-tidied format to preserve the integrity of the women's account.

Structure and Content

Across the interviews the women talked about their perceptions of their husbands' availability. Although their descriptions of husbands' availability differed, all stories were bound by the common theme of PPD symptoms. Their stories portrayed two kinds of husband availability: 'doing for' and/or 'being with.'

Their PPD Reality

The women's symptomatology experiences framed their stories about their husbands. As shown in Table 1, 2 and 3, the onset of their symptoms was associated with a specific timeframe. A number of women explicitly described their symptoms as "overwhelming," "unpredictable" and/or "incapacitating" in terms of the impact they had on their ability to meet their expectations of mothering:

M14: That night, it [the panic attack] happened again, but much worse.../I couldn't even talk on the phone.../cuz I didn't have the energy.../and all of a sudden I just couldn't calm down/and I was having a hard time breathing/um/my heart was racing/I got my husband/I told him "I think I'm dying"

M15: ... I didn't recognize depression. I grew up in a family that believed/pull up your socks and um/...I felt very isolated and really unhappy and then guilty about it. I remember that I was really worried, very worried/I had what turned out to be anxiety obsessional images of [child] that weren't so but I was so unstable and that happened so quickly

M19: And it just worsened and 3 days after I was home... I didn't want anything to do with him [the baby]/and he cried and cried, he had colic... I was on medication 3 days after the baby was born

M27: For that year [first year of her child's life] it was like I just wanted to be away from everybody and that's not like me... I just wanted to sleep and sleep and sleep/I was not the same person... I noticed the change in myself, my personality and my character/I just I didn't feel like I was/it felt like somebody took my body

M2: Like I can't feed [the child] properly, I can't change her properly/like I touch her she cries/I put her down she cries/like just somebody has to take care of [the child] because I can't... I was getting pretty frustrated with everything

M20: ...no matter how prepared you try to be you are never prepared enough/it is something that just happens and people get through it and that's it taken for granted... your world is just so distorted and I think that people around you are kinda going/"Oh my God" and they really don't know what to do either

M21: I wasn't myself/I wasn't acting like myself/I wasn't doing things I usually did/like it was all of a sudden/I didn't care anymore

Women often spoke about their difficulties in verbally articulating their suffering so that their husbands understood their needs. As M3 explained, "I don't even know how to explain it/it is hard to bring up the words." They acknowledged that difficulties in expressing themselves clearly contributed to family chaos and strained marital relations.

M9: ... and he wasn't hearing me/wouldn't listen to what I was saying/and he was walking away from me/and I grabbed it [a toy] and whack/so hard on my head that it split my head open/and blood starts coming down

M1: ... my husband was like "what the hell is wrong with you?" I said "you [my husband] have to call [someone for help]."... my husband was blind he really was/he was really blind to everything that was going on/and he was not a really big help in the beginning/I sort of felt like I was doing everything on my own you know

M14: ...I was thinking why is my husband planning for [the child's] future as an adult/and I was just constantly thinking that she was going to die because I was neglectful/or because I was doing something/that I ah/I in some ways/I thought/because I couldn't breastfeed I wasn't a good mother

M27: ... even my husband found it very hard to cause he's always tries to keep the peace like so it was very difficult for him to not upset me. It's just/I wasn't the same person

M1: So my husband... came out and said "well you know you're kind of not yourself"/and I was like "get lost"/... this was mid August/by September he was like/"no you need to go and see someone"

M3: ... it's always that my husband/it's always/he said the same thing all the time/"I can't read your mind"

In addition, some women did not initially interpret their experiences of fatigue, irritability, discomfort and/or tearfulness as PPD. Through their husbands' direct sharing of their observations, these women began to question if their inner psychological experiences were becoming apparent despite their efforts "to hide" symptoms or "not to tell" their husbands. Later, husbands confided that they sought validation of their genuine concern that "something was not quite right." Often they initiated consultations with immediate female family members. The husbands' worries were rooted in their knowledge of wife as a person.

M7: ... I told my husband about it [PPD] because he said/"you know/you're not/ you know you're changing/you've changed [a lot]"

M13: ... My husband involved my mother.../and actually he/he talked to my mother and said/"there's something wrong with her/this is not normal"/and my mother took him seriously and I really appreciated that cause ap-apparently he said I was at like three in the morning

M20: It is almost/are/there is a time when a mother is ready to hear/we need help... my husband telling me "there's something wrong/there is something wrong" but I'm like "no its your family, it's you"... there seems to be a certain time when the partner can say to you there's something wrong and you listen

M3: He [my husband] did notice that I was/there was something wrong with me but he couldn't see what it was/he didn't know about depression that much either/like he didn't have anything like that in his family so...

M4: My husband/he said there's something wrong/you need help.... so he did question but he himself didn't know where to go or what to do either

Husbands' Availability

The husbands' availability was of two types: 'doing for' and 'being with,' with the latter being the more typical mode. His physical presence, regardless, lessened their fear of "going crazy." Being "alone" in the home was perceived as begrudgingly tolerable, especially when husbands were contract or shift workers, requiring them to be out of the home for extended or unpredictable periods of time.

M8: I remember really well/I was picking [the baby] up from daycare at the end of the day and my husband I think was out of town overnight/so I mean I knew he wasn't going to be at the house that night/and I was just going to be on my own/and I was just...very overwhelmed

M14: The first few weeks when my husband went back to work I was afraid to be alone at night because that's when it [symptoms] got the worst

M18: ... there was something wrong/[my husband] went back to work and I couldn't be alone because I feared everything/so either I spent my days at my parents or my mom came for the day until [my husband] came back

M1: My husband worked the graveyard so I was alone all night with [the baby] and/and then he was sleeping during the day so I was alone with [the baby] all day

M22: My husband returned to working long hours/I was all alone/I never felt so helpless

M3: My husband did/he's the one that called the health unit/he's the one that found about the group/...before I used to be the one to call if there's anything

M24: It was just don't leave me alone because I was afraid... like I got overwhelming fear that I had [husband's name] go to the emergency with me and we talked to the doctor...

Additionally, their husband's physical presence was represented as an opportunity for women to verbally express their worries and uncertainties. Overall, the women were strategic about what, how and when they talked to their husbands. A criterion they used to judge the nature of their disclosure was their assessment of the potential risk or benefit. Women did not want to risk "further isolation," "frightening" their husbands or having their partners distance themselves by "never talking about what happened" in illness.

M6: I like/I was really pretty down at that time... I was really down and I wanted to talk to a friend who used to work in [social services]/I wanted to see her/the next day I get a call/he [caller from social service] said that the staff mentioned that I was pretty down and I wanted to go see Children's Aid/and that I should go/and then I told this to my husband/he's all worried about Children's Aid coming over/I said "no no"/I said I said/"I was having a bad day"/I guess that's why [my husband] freaked

M14: It's horrible/and so like I I had dreams that I was trying to do stuff to myself and my children/but I knew/and I brought it up to the doctor/he [the doctor] said/"if you have bad dreams/you need to get them out/whether it's to your husband"/or I didn't want to scare my husband more than I already had [laughter]

M3: I told my husband/I told him like I would dream of falling down the stairs/ so I break my leg or so I break my neck or something/it wouldn't be me doing it on purpose...why can't I just fall and die or something like/but/how can I even think that way? I couldn't/I know its illness

The above three examples about "telling my husband," however, did not include a description of husbands' apparent responses. Rather, within the context of their spousal relationship, the women implied that their verbal utterance of events or thoughts were "heard," thereby, reaffirming their interpretation. This partly may be associated with women's perceptions that their husbands "didn't know what to say/think or do." The following excerpts illustrate how these women connected with their husbands particularly when their mothering competency was questioned by others.

M6: I would go to the mall often/I just needed to be out and... [my friend] said to me "well your baby [at 3 months old] is lonely too/she needs more than to be a mall baby"/and that really bothered me/because I said to my husband "how did/ can [my friend] judge me like that?" "How did [my friend] know my baby was lonely?".../he [my husband] just/he just...

M17: /um/and at that point I realized okay you know/it's not something physical/it has to be emotional/so the realization came just mainly by myself and just telling my husband

'Doing For'

'Doing for' stories involved the husbands' response to the immediate needs of the family, both spouse and child. Circumstances compounded by PPD placed husbands in the challenging position of simultaneously being both family provider and carer.

M13: ... my husband couldn't get the/more time/well he was offered more time off work but there was no money there and we need to bring a pay check in pay for the house/pay for the house/all these other things

M17: ... bringing in take-out once in a while or/you know/just taking the baby out so I can have some alone time/um/picking her up from the babysitters so that helps to/so a lot of just little things that amount to a lot

M5: ... my husband with the baby at home and he's saying this isn't going away/ you need help you know/so he when in to talk to [his physician]/[the physician] told him that I have to come in/he took me in the next day and that's when the referral started happening/... it really helped with [partner's name] being able to change the diaper/make bottles

M20: lots of time I would tell him to do certain things and he'd do it

M26: he was supportive by taking any kind of job because of our financial stress

Frequently, in relation to their perceptions of "losing control," women explained how their husbands undertook care tasks to mediate symptoms. For example, they called substitute care providers, arranged child care, administered medications, or made themselves available to transport the women to healthcare providers.

M3: ... I couldn't ask for help for babysitting/for anything/he [my husband] had to/be- because to me if somebody says "no I can't tonight"/then it was like/ it's rejection

M13: ... my husband pretty much had to force the pills down my mouth/cause I was just like/"I'm not taking this I'm not that that I don't need this" and he'd like say "take your pill" and I would hate him for it

M14: um/I had to call a friend/my husband had to call a friend over/a friend I know that works for the [a health care agency]/and she came over while my husband could attend to the baby/and my other daughter/um/my husband or her had to help me nurse the baby

M16: ... husband encouraged me actually to call [a community support service] when I told him about it/he said/"well why don't you get out/get out of the house and call"/and yeah well I'm dealing with it and he's like/"no go and do this"/so he encouraged me

'Being With'

'Being with,' stories involved the husbands' affective presence that also promoted the women's sense of security. 'Being with' their husbands afforded women the opportunity to discuss their worries without fear of alienation. They perceived their husbands as respectfully close. Some women described their husbands' repeated offers to mutually share the day-to-day responsibilities as husband and wife.

> M17: [my husband's] been wanting to talk about it like asking me how I'm doing/and you know/"what are you thinking about"/and he's very tuned into my feelings so sometimes he'll know "okay what's going on?" "what are you thinking about?"

> M18: I finally broke down and told my husband there was something wrong/ so I kinda let him help/we did almost everything because I feared everything/my husband/um a lot of his care/I was able to talk with my husband mostly

> M25: I don't get along with my family/my husband would run interference by telling my mother it wasn't a good time to come over/making sure people who upset me stayed away

> M20: ... there seems to be a certain time when the husband can say to you/I'm frustrated and I don't know what to do

> M4: ...like [name of partner] being there and him learning a lot about the stuff that was happening

Aware of the impact of their symptoms, women gave their husbands permission to be their co-voice with trusted others such as familiar healthcare providers. In these cases, 'being with' involved his responsibility of interpreting or decoding the situation for their wives or families following a joint discussion.

> M8: I made my husband read one of those checklists about your feelings/and I said "I'm not gonna know if this happens to me/so you've got to look for these things"

> M14: And um, the next day my son had an appointment with the doctor/my family doctor/so when I went/um my husband and I explained what happened... one of my relatives said um to my husband/"oh/I seen a commercial about this postpartum depression" and "oh/[my husband] you know what/she doesn't catch that" and he is like "she does have it/you don't catch it"/you know/"it's um/it's hormonal/it's not about not managing having two kids"

> M1: ... [my husband] would let me know that I was getting better and he just tell me/"it's ok"

> M2:...like my husband and my relationship is way better although I feel guilt on a lot of fronts/I mean we've we've come from so far

> M3: We'd had to figure out a plan of attack we are going to take

M10: We were in such a strange place/we wanted our privacy... we called [a community service] because we didn't know where to start

Husbands were presented as responsible for preserving their wives' as mothers with external family members and, in limited cases, with a close family friend. Again, the public image 'as mother' was often a product of joint discussions bound by the family's privacy and cultural traditions.

M13: I had my husband and my mother.../he'd say [to my mother]/"she's pacing"/and she's like/my husband brought my mother just so that he would have the nerves to tell the doctor what he was seeing with me

M1: I told my husband to call my good friend...my husband called her in the middle of the night and told her what was going on/and she said "you gotta take her to the hospital"

M19: ... my husband and I explained to her [health care provider] what was going on and she's the one that said/"yeah/probably postpartum depression"...um restarting my relationship again with my husband

Meaning and Function

The meaning statements within stories address the question, "Why are the stories told?" The meaning of the availability stories was that partner as husband (and to a lesser extent father) was needed by the women who were struggling with motherhood in the context of PPD. The inclusion of messages such as "what can I [as husband] do to help?" "I told him," "I was heard," reflected women's perceptions of their husbands' tangible and intangible central involvement. During the period of adjustment following childbirth, 'doing for' and 'being with' stories revealed types of husbands' availability—physically, to do; affectively, to comfort, and cognitively, to interpret—each of these modes presented as integral to their integrity as women and shaped by the context of mothering.

Familiar patterns of availability seemed compromised during this stressful transition and compounded by the presence of PPD. The women's stories illustrate perceptions of the intrinsic care provided by husbands. Husbands' engagement in different modes of availability and their connectedness with their wives as trusted other seemed particularly adaptive. Variability in availability seemed to be valued given the mothers' description of the ambiguity, unpredictability and unexpectedness of PPD. In retrospect, husbands communicated their availability through expressions such as "its okay," or through actions such as taking the day off work, or even when he was directive in stating, "take your pill." Participants evaluated the husbands' availability as purposefully caring for them as women.

Case Example

Nancy (M11) was a young first-time mother, on maternity leave from her professional occupation. She lived with her husband and infant in a residential neighborhood. Her husband's employment situation was tenuous given that was a contract position in the service industry. Her own mother did not live in the region and was not accessible for support after the birth of Nancy's child due to commitments within the extended family. Her mother-in-law resided in the area and was able to help if asked. She described the postpartum period as a time when "my life was extremely scary." Her story included the emotional, physical, social and functioning implications typically associated with PPD. Nancy's desire was "to be the best mom that was possible." Thereby, she recounts her preference to be independently responsible for all the child's nurturing, so that he "could sleep" and continue to fulfill his obligations as provider for the family. Within two months, however, Nancy realized that she was "completely lost" and knew, "in [her] heart" that she "needed" her husband to become more involved so that she could attend to her increasing mental health needs. The following story illustrates an aspect of Nancy's husband's availability. It is presented using Labov and Waletzky's [49,57] structural elements to make the analysis process transparent. Each line of the dialogue is numbered and represents her interaction with the researcher including speech interruptions, false starts, and overlaps in speaking turns. This story was prompted by the researcher's inquiry and in turn, Nancy's response without further dialogue from the researcher.

Story Stimulus:

533 Researcher: To clarify, a community nurse came to visit?

Orientation:

534 M11: I had a visit from a nurse/and that was it/I had a visit

535 /someone came in and weighed the baby/they weighed the baby/she

536 did this to see if the baby was drinking?

Complicating Action:

537 She [nurse] asked me "are you breastfeeding?"

Evaluation/Complicating Action

538 "Is any milk coming in?"/she kept focusing on the breastfeeding part

539 you know/at this time like/put on your smiley face

Complicating Action:

540 umm I wanted to tell her "listen/it so hard..."

Resolution:

541 I told my husband after that

Evaluation:

542 /I can remember how this visit made me feel bad/it made me guilty

Nancy's reference to her husband after experiencing an unsettling mothering event seems tangential because she does not verbalize his immediate response. Her chosen disclosure of this personal experience to him as husband infers her perception of his availability as a safe and protective other when she herself has doubts about her adequacy of mother. Nancy's transcript continues uninterrupted with the immediate introduction of another story that emphasizes her husband's availability.

Orientation

543 cause when my mom was here I was upstairs pumping for two
544 hours to get two ounces

Complicating Action:

545/the nurse reminded me, "it's not that hard/come on I'll help"/the
546 baby is screaming like blue murder

Evaluation:

547/and the nurse goes, "it is not working"/I wanted to scream

Orientation/Complicating Action:

548 I'd remember when my mother-in-law come to visit/running up stairs 549 and pump pump pump pump pump cause [child's name] needs a
550 feeding/this is what is best for the baby

Abstract:

551 I'm a poor mother because I can't do this/I wasn't able to breastfeed

Evaluation/Complicating Action/Evaluation:

552/I bought formula but/

553 whenever you're in that state of mind/I couldn't/it was so painful

554 /cry, cry cry because not even a drop was coming out

555 and my mom said/"listen your kid is not gonna die"

556/my husband said he was raised on carnation milk/honey and water 557/and he's like normal functioning

Nancy expended much time, energy, and pain to fulfill her expected role as the "best" mother including breastfeeding. Despite all her intended efforts to breast-feed, and solicited consultations with others (nurse, mother, and mother-in-law), it was her husband's verbal response and affective presence that reassured her and lessened her guilt as a mother. In this case study, Nancy's husband was needed as she struggled to meet her ideals as a mother that were reinforced and advocated by others. Nancy's matter-of-fact recounting of her husband's seemingly purposeful and timely availability was interpreted as caring for her.

Discussion

The initial focus on maternal help-seeking practices in PPD in the larger study generated this study's data subset suggesting the essential availability of husbands for women experiencing PPD. As evidenced in their stories, for this group of women, the availability or presence of "husband" was largely perceived as provid-ing affirmation and security. Consistent with other research [8], these women's experience in PPD was destabilizing, and negatively interfered with their ability and desire to attend to their roles as wife and mother. To establish balance, the women required their husbands' physical, affective and cognitive involvement as they struggled with PPD. The interpretation of the husband's availability ranged from conflicting, reassuring and/or nurturing in relation to the woman's perceived health status and efforts to mother.

Although men generally desire to be supportive to their partners, the demands of infant care, even in the absence of PPD, have been found to contribute to rela-tional adjustments and strain [60]. In the postpartum period, Blum [61] proposed that the availability of a sympathetic other counterbalances another's inherent dependency needs, whether verbalized or not. Women's risk for maternal distress increased when they find no one available to assist them in meeting such rudimentary psychodynamic needs. Basic attentive listening to a mother's concerns may address her needs without requiring her "to acknowledge them any more than [she] can." [61] The current study's findings suggest that over the course of PPD a husband's availability may need to modulate from a physical presence to a reciprocal exchange. Such unpredictable and fragile transitions appeared to communicate commitment to, responsibility for and preservation of the woman and the family. Although not explored in this study, previous research suggests that men experience their own struggles with maternal PPD while intending to lessen the suffering of their partners through attempts at involvement [4,5,32,33].

For these women, understandings of their husbands' physical availability may have been as subtle as their "sitting in the chair" or "just being in the house." Potentially, husband's physical proximity permitted a connection, even an opportunity for him to observe and protect his wife. Consistent with the findings of other studies [62], several of this study's participants did not initially view themselves as ill. Some women in this study engaged in self-silencing, an interpersonal style intended to minimize their partners' tension, worry or rejection. As a result, his availability and intentionally cautious watching because of unfamiliar circumstances may be considered a positive source of support and validation [63,64]. In addition, husbands' verbalization of their perceptions about their wives' behaviors and challenges in attending to the needs of their infants was critical in legitimizing the women's distress and assisting in seeking help.

This study's group of women reflected on the importance of 'doing for,' or instrumental support as identified in other studies [1,24,62]. Because of the women's vulnerability, 'doing for' appeared to be a one-way transaction aimed at performing a task such as driving, cleaning or administering medication. Consistent with other studies [4], emphasis was on what the husband did to "fix" emergent problems; helping to mobilize resources to meet his partner's or family's needs. For the dependence/interdependence health of the couple, our findings reinforce the importance of a network of resources to moderate the health risks for each partner. This dynamic is reinforced in Goodman's [11] integrative review of PPD in fathers; depression in one partner influences the health of the other.

Barclay and Lupton [33] found that, for men, "being there" refers to a physical and emotional sensitivity for the sake of family stability and mutual support. Although many of the men in their study remained on the "fringes of parenthood," for the first sixth months if their child's life, they were committed to providing responsive care for their partners' and infants' needs. As Goodman suggests [60], despite fathers' intentions to "be there," the partners' redefinition of roles is characterized as a process of "trial and error." From the perspectives of women in the present study, husbands' availability, as part of the transition, afforded the mothers the opportunity to verbalize select feelings and thoughts, thereby possibly lessening inner emotional chaos.

For this group of women, 'being with' their husbands was expressed in less tangible terms than their husbands' 'doing for' pattern of availability. These narratives do suggest that their spousal relationship provided the context for reciprocity as the husband listened to, advocated for, and negotiated with his partner. As Doucet [12] proposed, "being with" is an inter-subjective connection: partners interacting, learning to move through spaces such as parenthood congruent with public expectations concurrently making judgments about maintaining or avoiding social encounters. In terms of the buffering effects of spousal support in PPD, women who perceive their partners as emotionally supportive reported less avoidance behavior, greater martial satisfaction and less depressive symptomatology [1,16,65]. The combination of these 'doing for' and 'being with' results support Ontario, Canada's recent family development services directed at both adults sharing responsibility for nurturing intra-familial environment during transitions.

With regard to secondary prevention, research has demonstrated that postpartum programs emphasizing relational interventions were associated with positive maternal and family outcomes [18]. Paris and Dubus [17] explored volunteer, paraprofessional and home visitors' relational interventions with postpartum families. Home visitors perceived that their provision of instructional and emotional support, validation and affirmation positively influenced the development of partner and parent-child relationships. Although the current study findings do not advocate for care within the marital relationship to become professionalized, women recommended the need for information and skills that focus on cultivating mutually empathic relationships. Future longitudinal research should explore men's perceptions of their informal care giving processes in PPD.

The researchers of this study have not attempted to make claims of causation or truth [58,66]. Instead, through repeated engagement with the data an understanding from women's representations of their husbands' central involvement in PPD has been constructed. Several of the women who self-identified for this study had previously shared their PPD experiences through informal peer support groups, potentially influencing the results. These women may have processed a more thoughtful view of their husbands as a result of the positive effect of their support-seeking behaviors. As previously indicated, to enhance the study's trustworthiness a research team member who was a PPD peer support worker and, therefore, contextually aware, assisted the academic researchers to "see what [they] could] not yet see" [35]. Further, to address representative credibility [33] in relation to women's help seeking in PPD while ensuring their confidentiality, the lack of a standardized collection of particular identifying socio-demographic information from women living in smaller communities may limit the transferability of the findings. As Thorne [33] suggests, however, we were aware that our chosen methods had to reflect rather than define our research purpose.

Conclusion

This study described a sample of 27 women who self-identified as having PPD and their understanding of their husbands' availability. During this time of vulnerability, their husbands' physical, emotional and cognitive availability positively contributed to their functioning and self-appraisals as wife and mother. Their representations of their husbands' 'doing for' and/or 'being with' promoted their well-being and ultimately protected the family. Given that husband is perceived as central in mitigating women's suffering with PPD, the consistent implementation of a triad orientation that includes wife, partner and child as opposed to a more traditional and convenient dyadic orientation is warranted in comprehensive postpartum care. Finally, this study contributes a theoretical understanding of responsive as well as reactive connections between a couple during the postpartum period.

Abbreviations

(DSM-IV-TR): Diagnostic and Statistical Manual of Mental Disorders, 4th Edition, Text Revision; (PPD): Postpartum depression.

Competing Interests

The authors declare that they have no competing interests.

Authors' Contributions

All authors contributed to the research's conception and design. The manuscript was written by PM and PB. The other authors read and approved the final manuscript.

Acknowledgements

This study was funded by the Louise Picard Public Health Research Grant (formerly the Laurentian University-Sudbury and District Health Unit Fund) and a SSHRC SIG Grant. The authors gratefully acknowledge the assistance Anna

Love, a Year IV nursing student, for her assistance with this manuscript, and Professor Sharolyn Mossey for her critically review of this manuscript.

References

1. Dennis C, Ross L: Women's perceptions of partner support and conflict in the development of postpartum depressive symptoms. J Adv Nurs 2006, 56(6):588–599.

2. Beck CT: Predictors of postpartum depression. Nurs Res 2001, 50(5):275–285.

3. O'Hara MW, Swain AM: Rates and risk of postpartum depression-A meta-analysis. International Review of Psychiatry 1996, 8(1):37–54.

4. Davey SJ, Dziurawiec S, O'Brien-Malone A: Men's Voices: Postnatal Depression From the Perspective of Male Partners. Qual Health Res 2006, 16(2):206–220.

5. Tammentie T, Paavilainen E, Åstedt-Kurki P, Tarkka M: Family dynamics of postnatally depressed mothers - discrepancy between expectations and reality. J Clin Nurs 2004, 13(1):65–74.

6. Paley B, Cox MJ, Kanoy KW, Harter KSM, Burchinal M, Margand NA: Adult Attachment and Marital Interaction as Predictors of Whole Family Interactions During the Transition to Parenthood. Journal of Family Psychology 2005, 19(3):420–429.

7. Dressel PL, Clark A: A critical look at family care. Journal of Marriage & the Family 1990, 52(3):769–769.

8. Beck CT: Theoretical perspectives of postpartum depression and their treatment implications. MCN 2002, 27:282–287.

9. Holtslander L: Clinical Application of the 15-Minute Family Interview: Addressing the Needs of Postpartum Families. Journal of Family Nursing 2005, 11(1):5–18.

10. Morgan M, Matthey S, Barnett B, Richardson C: A group programme for postnatally distressed women and their partners. J Adv Nurs 1997, 26(5):913–920.

11. Goodman JH: Paternal postpartum depression, its relation to maternal postpartum depression, and implications for family health. Journal of Advanced Nursing 2004, 45(1):26–35.

12. Doucet A: 'Estrogen-filled worlds': fathers as primary caregivers and embodiment. Sociol Rev 2006, 54(4):696–716.

13. Fraser C, Warr DJ: Challenging Roles: Insights Into Issues for Men Caring for Family Members With Mental Illness. American Journal of Men's Health 2009, 3(1):36–49.

14. Page M, Wilhelm MS: Postpartum Daily Stress, Relationship Quality, and Depressive Symptoms. Contemporary Family Therapy: An International Journal 2007, 29(4):237–251.

15. Scrandis DA: Normalizing Postpartum Depressive Symptoms With Social Support. Journal of the American Psychiatric Nurses Association 2005, 11(4):223–223.

16. Bost KK, Cox MJ, Burchinal MR, Payne C: Structural and supportive changes in couples' family and friendship networks across the transition to parenthood. Journal of Marriage and Family 2002, 64(2):517–517.

17. Paris R, Dubus N: Staying Connected While Nurturing an Infant: A Challenge of New Motherhood*. Family Relations 2005, 54(1):72–83.

18. Registered Nurses' Association of Ontario (RNAO): Interventions in Postpartum Depression. 1st edition. Toronto, Canada: Registered Nurses' Association of Ontario; 2005.

19. Oates MR, Cox JL, Neema S, Asten P, Glangeaud-Freudenthal N, Figueiredo B, Gorman LL, Hacking S, Hirst E, Kammerer MH, Klier CM, Seneviratne G, Smith M, Sutter-Dallay A-, Valoriani V, Wickberg B, Yoshida K, TCS-PND Group: Postnatal depression across countries and cultures: A qualitative study. British Journal of Psychiatry. Special Issue: Transcultural Study of Postnatal Depression (TCS-PND): Development and testing of harmonised research methods 2004, 184(46):s10-s10.

20. American Psychiatric Association: Diagnostic and Statistical Manual of Mental Disorder. 4th edition. Washingtion, DC: American Psychiatric Association; 2000.

21. Rodrigues M, Patel V, Jaswal S, de Souza N: Listening to mothers: qualitative studies on motherhood and depression from Goa, India. Soc Sci Med 2000, 57(10):1797–1806.

22. Malik NM, Boris NW, Heller SS, Harden BJ, Squires J, Chazan-Cohen R, Beeber LS, Kaczynski KJ: Risk for maternal depression and child aggression in Early Head Start families: A test of ecological models. Infant Mental Health Journal. Special Issue: Infant Mental Health in Early Head Start 2007, 28(2):171–171.

23. Misri S, Kostaras X, Fox D, Kostaras D: The Impact of Partner Support in the Treatment of Postpartum Depression. Canadian Journal of Psychiatry 2000, 45(6):554–558.

24. Smith LE, Howard KS, Centers for the Prevention of Child Neglect, US: Continuity of paternal social support and depressive symptoms among new mothers. Journal of Family Psychology 2008, 22(5):763–773.

25. Bell L, Goulet C, St-Cyr Tribble D, Paul D, Boisclair A, Tronick EZ: Mothers' and Fathers' Views of the Interdependence of Their Relationships With Their Infant: A Systems Perspective on Early Family Relationships. Journal of Family Nursing 2007, 13(2):179–200.

26. Hall P: Mothers' experiences of postnatal depression: an interpretative phenomenological analysis. COMMUNITY PRACT 2006, 79(8):256–260.

27. Holopainen D: The experience of seeking help for postnatal depression. Aust J Adv Nurs 2002, 19(3):39–44.

28. Everingham CR, Heading G, Connor L: Couples' experiences of postnatal depression: A framing analysis of cultural identity, gender and communication. Soc Sci Med 2006, 62(7):1745–1756.

29. Tammentie T, Tarkka M, Åstedt-Kurki P, Paavilainen E, Laippala P: Family dynamics and postnatal depression. J Psychiatr Ment Health Nurs 2004, 11(2):141–149.

30. Webster A: The forgotten father: the effect on men when partners have PND. COMMUNITY PRACT 2002, 75(10):390–393.

31. Boath EH, Pryce A, Cox J: Postnatal depression: The impact on the family. J REPROD INFANT PSYCHOL 1998, 16(2):199–199.

32. Meighan M, Davis MW, Thomas SP, Droppleman PG: Living with postpartum depression: the father's experience. MCN 1999, 24(4):202–208.

33. Barclay L, Lupton D: The experiences of new fatherhood: a socio-cultural analysis. J Adv Nurs 1999, 29(4):1013–1020.

34. Thorne S: Peals, pith and provocation; ethical and representational issue in qualitative secondary analysis. Qualitative Health Research 1998, 8:547–555.

35. Thorne SE: Interpretive description. Walnut Creek, CA: Left Coast Press; 2008.

36. Thorne S, Kirkham SR, MacDonald-Emes J: Focus on qualitative methods. Interpretive description: a noncategorical qualitative alternative for developing nursing knowledge. Res Nurs Health 1997, 20(2):169–177.

37. Statistics Canada [http:/ / www12.statcan.ca.librweb.laurentian.ca/ census-recensement/ 2006/ dp-pd/ prof/ 92-591/ details/ Page.cfm?Lang=E&Geo1=CSD&Code1=3553 005&Geo2=PR&Code2=35&Data=Count&Sea rchText=sudbury&SearchType= Begins&S earchPR=01&B1 =All&Custom=].

38. North East LHIN [http:/ / www.nelhin.on.ca/ assets/ 0/ 16/ 42/ 1228/ ea467f0814e640169c799b41936141db.pd f]

39. May KA: Interview techniques in qualitative research: Concerns and challenges. In Qualitative nursing research: A contemporary dialogue (rev. ed.). Edited by: Morse JM. Thousand Oaks, CA, US: Sage Publications, Inc; 1991:188–201.

40. Mishler EG: Research Interviewing: Context and Narrative. Cambridge: Harvard University Press; 1986.

41. Warren CAB: Qualitative Interviewing. In Handbook of interview research: Context and method. Edited by: Gubrium JF, Holstein JA. Thousand Oaks, California: Sage Publishers, Inc; 2001.

42. Bruner J: The Narrative Construction of Reality. Critical Inquiry 1991, 18(1):1–21.

43. Clandinin DJ, Connelly FM: Personal experience methods. 1994, 413–427.

44. Coffey A, Atkinson P: Making Sense of Qualitative Data: Complementary Research Strategies. 6th edition. London: Sage Ltd; 1996.

45. Connelly FM, Clandinin DJ: On Narrative Method, Personal Philosophy, and Narrative Unities in the Story of Teaching. Journal of Research in Science Teaching 1986, 23(4):293–310.

46. Langellier KM: Personal Narratives: Perspectives On Theory and Research. Text & Performance Quarterly 1989, 9(4):243.

47. Mattingly C, Garro LC: Narrative representations of illness and healing: Introduction. Social Science & Medicine 1994, 38:771–774.

48. Riessman CK: Narrative analysis. Thousand Oaks, CA, US: Sage Publications, Inc; 1993.

49. Labov W, Waletzky J: Narrative analysis: oral versions of personal experience. In Essays on the Verbal and Visual Arts. Edited by: Helms J. Seattle: University of Washington Press; 1972:12–44.

50. Agar M, Hobbs J: Interpreting discourse: coherence and the analysis of ethnographic interviews. Discourse Processes 1982, 5:1–32.

51. Bailey PH: Death stories: acute exacerbations of chronic obstructive pulmonary disease. Qual Health Res 2001, 11(3):322–338.

52. Bailey PH: The Dyspnea-Anxiety-Dyspnea Cycle--COPD Patients' Stories of Breathlessness: "It's Scary/When You Can't Breathe.." Qual Health Res 2004, 14(6):760–778.

53. Bailey PH, Tilley S: Storytelling and the interpretation of meaning in qualitative research. J Adv Nurs 2002, 38(6):574–583.

54. Bailey PH, Colella T, Mossey S: COPD-intuition or template: nurses' stories of acute exacerbations of chronic obstructive pulmonary disease. J Clin Nurs 2004, 13(6):756–764.

55. Bennett G: Narrative as Expository Discourse. The Journal of American Folklore 1986, 99(394):415–434.

56. Antaki C: Exploring and Arguing: The Social Organizations of Accounts. London: Sage Publications; 1994.

57. Labov W, Waletzky J: Narrative Analysis: Oral versions of personal experience. Journal of Narrative and Life History 1997, 7:3–38.

58. Bailey PH: Assuring Quality in Narrative Analysis. West J Nurs Res 1996, 18(2):186.

59. Cutcliffe JR, McKenna HP: When do we know that we know? Considering the truth of research findings and the craft of qualitative research. Int J Nurs Stud 2002, 39(6):611–618.

60. Goodman JH: Becoming an involved father of an infant. JOGNN: Journal of Obstetric, Gynecologic, & Neonatal Nursing 2005, 32(2):190–200.

61. Blum LD: Psychodynamics of Postpartum Depression. Psychoanalytic Psychology 2007, 24(1):45–62.

62. Letourneau N, Duffett-Leger L, Stewart M, Hegadoren K, Dennis C, Rinaldi CM, Stoppard J: Canadian mothers' perceived support needs during postpartum depression. JOGNN 2007, 36(5):441–449.

63. Henderson J: 'He's not my carer--he's my husband': personal and policy constructions of care in mental health. Journal of Social Work Practice 2001, 15(2):149–159.

64. Whiffen VE, Kallos-Lilly AV, MacDonald BJ: Depression and attachment in couples. Cognitive Therapy and Research 2001, 25(5):577–577.

65. Lemola S, Stadlmayr W, Grob A: Maternal adjustment five months after birth: the impact of the subjective experience of childbirth and emotional support from the partner. J REPROD INFANT PSYCHOL 2007, 25(3):190-202.

66. Paley J, Eva G: Narrative vigilance: the analysis of stories in health care. Nursing Philosophy 2005, 6(2):83–97.

Challenges at Work and Financial Rewards to Stimulate Longer Workforce Participation

Karin I. Proper, Dorly J. H. Deeg and Allard J. van der Beek

ABSTRACT

Background

Because of the demographic changes, appropriate measures are needed to prevent early exit from work and to encourage workers to prolong their working life. To date, few studies have been performed on the factors motivating continuing to work after the official age of retirement. In addition, most of those studies were based on quantitative data. The aims of this study were to examine, using both quantitative and qualitative data: (1) the reasons for voluntary early retirement; (2) the reasons for continuing working life after the official retirement age; and (3) the predictive value of the reasons mentioned.

Methods

Quantitative data analyses were performed with a prospective cohort among persons aged 55 years and older. Moreover, qualitative data were derived from interviews with workers together with discussions from a workshop among occupational physicians and employers.

Results

Results showed that the presence of challenging work was among the most important reasons for not taking early retirement. In addition, this motive appeared to positively predict working status after three years. The financial advantages of working and the maintenance of social contacts were the reasons reported most frequently for not taking full retirement, with the financial aspect being a reasonably good predictor for working status after three years. From the interviews and the workshop, five themes were identified as important motives to prolong working life: challenges at work, social contacts, reward and appreciation, health, and competencies and skills. Further, it was brought forward that each stakeholder can and should contribute to the maintenance of a healthy and motivated ageing workforce.

Conclusion

Based on the findings, it was concluded that measures that promote challenges at work, together with financial stimuli, seem to be promising in order to prolong workforce participation.

Background

One of the most notable changes in the working population is its ageing. The baby boom cohorts born after the Second World War (born between 1945 and 1965) are now middle-aged; the oldest of them have already started retiring. At the same time, lower birth rates in the past few decades imply that fewer young people will be entering the labour market [1]. These demographic changes are bringing about a shift in the ratio of workers to retirees that will lead to a relative shortage of active workers.

Of the major regions of the world, the process of population ageing is most advanced in Europe [2]. The median age of European Union (EU) citizens will increase between 2004 and 2050 from 39 to 49 years [3]. After 2010, the year that will mark the greatest number of members of the potential working population (i.e. those between 15 and 64 years), the population of working age will decline from 331 million to 268 million in 2050 [3]. In contrast, the proportion of people 65 years and older will increase.

These two demographic changes will result in an increase in the old-age dependency ratio (i.e. the number of people over 65 divided by the number of working-age people) from 25% today to about 53% in 2050 for the 25 EU countries [3]. At the same time, the share of older workers (i.e. those between 55 and 64 years) in the total potential workforce will logically increase. It is estimated that by the year 2025, between one in five and one in three workers will be an older worker [2].

It is clear that the demographic shift has serious economic and social implications, among others the financing of the social security systems, in that a shrinking number of economically active people (i.e. workers) will have to pay for the national pensions of an increasing number of retired persons. The ageing of the workforce also implies a change in the human resources (HR) strategies to manage age in the workplace. Thus, both government and private companies face the challenge of finding means to prolong the labour participation of (older) workers. This is especially true since, to date, many older people have left their jobs either voluntarily (i.e. because of early retirement) or involuntarily (i.e. because of work disability or unemployment) much earlier than the official age of retirement [4].

As in many countries, the social security system in The Netherlands used to offer the opportunity of retiring with a pension before the official retirement age of 65. This so-called early retirement pension (ERP) was implemented during a period of widespread unemployment, with the intention of providing better opportunities for the younger generation to find jobs. However, due to the population's ageing and its consequences, these early exits from work are no longer affordable from an economic point of view. Instead, appropriate measures are needed to prevent early exit from work and to encourage workers to prolong their working life.

During the last few years, measures discouraging early retirement have been implemented in many countries worldwide. For example, since 2006, ERP is no longer supported fiscally in the Netherlands, so that voluntary early exit from work has become more expensive. In other countries, raising the mandatory retirement age is one of the measures implemented. It may, however, be questioned whether such measures imposed by the government or the employer are effective. Commitment from the target group, i.e. the (ageing) workforce, is an important aspect for successful implementation.

To date, most of the research has focused on the determinants of early exit from work [5,6]. As far as the authors are aware, there are only limited data as to the motives of employees for prolonging working life. For example, a study of Lund and Borg [7] showed that very good self-rated health and high development possibilities were independent predictors for remaining at work among males. Among females, the same two predictors were found in addition to high decision authority, medium-level social support and absence of musculoskeletal problems in the knees [7].

In addition, some other recent studies showed that retirement decisions are influenced not only by the worker's health status, but also by income levels and pension rights [8,9]. Those aged 50 and over with poor health, high income or accumulated wealth and access to occupational pensions are more likely to retire at the normal retirement age or retire early [8,9]. Another study showed domestic and family considerations to be important influences of retirement behaviour [10]. In contrast, the evidence about the determinants of involuntary exit from work due to work disability shows occupational factors to be among the most important determinants [11,12].

However, the evidence as to the reasons for voluntary early exit from work is more scarce. From the few previous studies, it can be concluded that retirement decisions on a voluntary basis are multidimensional and not driven by any one single factor. In addition, the little available evidence as to the reasons for voluntary early retirement as well as for continuing working life is based mostly on quantitative data. There have been only a few attempts that involved qualitative data incorporating the worker's opinions about the factors that motivate them to prolong their work career after the official age of retirement.

For example, a semistructured interview study among persons who chose early retirement and those who did not, supported the quantitative finding that decisions about retirement are not made in a vacuum, but have to do with diverse types of possible routes into retirement [13]. These dealt with organisational restructuring, financial offers and opportunities for leisure and self-fulfilment that early retirement offers [13].

From a second qualitative study, it appeared that the way of conceiving work and retirement varied among persons from different socioeconomic backgrounds [14]. The conclusion was that various factors, including financial imperatives and HR practices, intersect at state pension age to shape people's routes into retirement and their options for continuing in work [14].

Finally, most of the previous studies have used cross-sectional data. Hence, the predictive value of the motives to retire early mentioned by those still working remains unclear.

Based on the limited literature on the determinants for prolonging working life, and the scarcity of qualitative data, the aims of the present study were: (1) to examine the reasons for voluntary early retirement; (2) to examine the motives for continuing working life after the official retirement age; and (3) to examine the predictive value of the reasons mentioned. A mixed-methods approach was applied, with quantitative and qualitative data.

Methods

This article describes the results of three studies. The first study includes data analyses of a prospective study among persons aged 55 years and older. The second study is based on qualitative data from interviews with workers, while the third study includes a workshop among occupational physicians (OPs) and employers.

Study 1. Quantitative Study (LASA)

The aim of this quantitative study was to examine the reasons for voluntary early retirement (first study aim) as well as the reasons not to voluntarily retire early. Moreover, with the data of both baseline and follow-up (i.e. three years later), the predictive value of the motives stated at baseline was determined (third aim).

Study Sample

The first study sample consisted of participants of the Longitudinal Aging Study Amsterdam (LASA). LASA is an ongoing, multidisciplinary, cohort study among persons aged 55 and over. It focuses primarily on the predictors and consequences of changes in older persons' physical, cognitive, emotional and social functioning.

The sampling and data collection procedures and the response rates were described in detail elsewhere [15,16]. In summary, LASA started with data collection in 1992–1993. A random sample of persons aged 55 years and over (birth years between 1908 and 1937), stratified by age, sex, urbanisation grade and expected five-year mortality was drawn from the population registers of 11 municipalities in three regions in The Netherlands. This procedure led to a representative sample of the Dutch older population, reflecting the national distribution of urbanisation and population density.

In 2002–2003, a second sample of men and women aged 55 to 64 years was drawn with the same sampling frame as the original cohort. The 2002–2003 sample is the sample for the current study, comprising 1002 participants aged 55–64 years (response rate 57%). In 2005–2006, a follow-up measurement of the second cohort took place (n = 908).

Written informed consent was obtained from all participants. The study was approved by the Medical Ethics Committee of the VU University Medical Centre.

Interviews

The interviews were held at the respondents' residences and were conducted by trained interviewers, who used laptop computers for data entry. The structured interview covered a wide range of topics related to physical and cognitive health and social and psychological functioning. For the purpose of the present study, the interview also included questions on reasons for considering early retirement.

The respondents were asked several questions as to their working status. Questions relevant to this study were: (1) Are you currently working in a paid job? (yes; no); (2) Are you currently (partially) work-disabled? (yes; no); (3) Have you already taken (partial) early retirement? (yes, completely; yes, in part; no); and (4) Would you consider taking (partial) early retirement if financially possible? (yes; no). Partial early retirement refers to working fewer hours in the main occupation.

To get insight into the reasons for voluntary early retirement as well as the reasons not to voluntarily retire early, respondents were asked their most important reason: (1) to take (partial) early retirement; (2) not to take full early retirement; (3) not to take early retirement at all; and (4) (among those who had already taken (partial) early retirement) to have taken (partial) early retirement. All four questions included branching questions that were asked to subgroups according to the working status and the consideration to take (partial) early retirement (Figure 1). The first two questions were asked of those with a paid job, who had not already taken early retirement, but who were considering taking early retirement, whereas the third question was asked of those with a paid job, who had not already taken full retirement and who were not considering taking early retirement. For each of these questions, the last of the five or seven answer categories included "another reason" than mentioned, leaving respondents space to fill in their own reason.

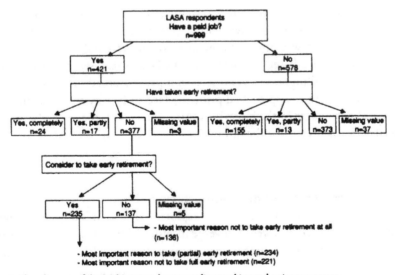

Figure 1. Flow diagram of the LASA respondents regarding working and retirement status.

Analysis

For the purposes of this study, descriptive analyses were conducted. A frequency table was produced for each of the four questions (see above) indicating the percentage of each reason specified. The "other reason" category was analysed in more detail; answers that could be clustered were grouped. The predictive value of the reasons reported was determined by a frequency table of the working status (i.e. working or having retired) at follow-up per reason mentioned at baseline. Analyses were performed using SPSS software, version 14.0 for Windows.

Both Study 2 and Study 3 were performed to get insight into the motivating factors for continuing to work and the measures that can be taken to stimulate prolonging working life.

Study 2. Interviews with Workers

Study Sample

Workers were recruited by the occupational health service (OHS) that participated in this study. By means of its customer database, the OHS approached 12 companies that differed in sector and size. Four companies agreed to participate. These four companies were from different sectors and included: (1) local government, (2) youth- and health care, (3) outdoor advertising and (4) an OHS, located in another city than the one involved in the recruitment in this study. The companies also varied in size (from approximately 60 to 700 workers), job characteristics and workers' educational level.

Within each company, the aim was to interview two workers individually and to hold one focus group interview with approximately five to 10 workers. Selection of the workers was done by a member of the HR staff or the head of the department, and was based on socioeconomic factors (e.g. sex, age, job position), availability and willingness to volunteer, to capture a broad range of characteristics. The workers were approached primarily face-to-face by their HR staff or supervisor to participate in the interviews. After agreement, the researcher arranged a specific date and time for the interviews.

Interviews

For the purposes of this study, semistructured interviews were held. Semistructured interviews define the area to be explored, at least initially, and allow the interviewer or the interviewee to diverge in order to pursue an idea in more detail [17]. This strategy encourages open answers, thereby eliciting new, additional information. During the interview, the interviewer tried to be interactive and to uncover factors that were not anticipated at the outset of the interview.

After briefly introducing the study and asking a few general questions, the interview guide posed the following questions: (1) "Are you willing to continue working until or after the age of 65?" (2) "What are factors that motivate you to continue working?" (3) "What measures can or should be taken to prolong your work career?"

All interviews took place face-to-face and were recorded on a digital voice recorder. The focus group interviews lasted approximately 50 minutes; the duration of the individual interviews varied from 24 to 42 minutes. During the interviews, the interviewer took field notes. The interviews were held in a meeting room at the company and were conducted by the principal researcher (KP).

Analysis

The interviews were fully transcribed by an assistant. Subsequently, content analysis was conducted by the principal researcher to analyse the transcripts. First, the transcripts were read and reread to become familiar with the text. Next, the text was marked with codes indicating the content of the response. The codes were then grouped together into key themes. In the Results section, interviewees' quotations that were considered representative for the theme are reported in order to illustrate the meaning of the themes.

Study 3. Workshop with Occupational Physicians (OP) and Employers

OPs and Employers

To compare the views of the workers with the opinions of important stakeholders, a two-hour workshop among OPs and employers was held. The workshop was organised within a general course for OPs by their occupational health service (OHS). As the workshop fit well in the programme, it was decided to incorporate the workshop in the OHS's OP course. In total, 20 OPs participated in the course, including the workshop. In addition, five representatives (human resource management (HRM) staff) of the four participating companies joined during the workshop.

Workshop

The workshop started with a 30-minute presentation by the principal researcher about the study and the results of the interviews among the workers. Subsequently, four working groups were formed, each consisting of one or two representatives of each company and five OPs.

For one hour, each working group discussed two issues. First, they discussed the motivating factors mentioned by the workers. In their discussion, the OPs and employers were encouraged to add motivating factors. The second issue discussed in the working group concerned the measures to be taken by the employer or the OHS that might stimulate workers to prolong their participation in the workforce. Each working group was asked to write down its views, and one person within each group was asked to report on these in the plenary session. In the plenary session, per working group, the workshop leader (KP) wrote down all views of each working group on a flip chart and gathered the papers of each working group.

Analysis

After the workshop, the views reported on the flip chart and the working groups' papers were copied by the researcher in an electronic form on a computer. The workshop notes were coded according to the themes identified by the interviews with workers. Similar to the analysis of the interviews (Study 2), the text was marked with codes and then grouped together into themes.

Results

Study 1

Table 1 shows the working status of the study population at baseline. A small majority (57.9%) did not have a paid job (any longer) at the moment of baseline measurement, and about a quarter of the respondents (23%) were work-disabled. The large majority (78.2%) had not taken early retirement. Of those currently working (n = 421), almost two thirds (63.2%) reported they were considering taking (partial) early retirement (Table 1). Further, among those with a paid job, n = 377 were not yet partially retired early (Figure 1).

Table 2 presents the frequencies of the workers' most important reasons not to take (full) early retirement. From this table, it can be concluded that the reasons for not taking early retirement at all are different from the reasons for not taking full early retirement. Having sufficient challenges at work appeared to be by far (59.6%) the most important reason for workers not to take early retirement, whereas the financial aspect (32.6%) and the social contacts (25.3%) were reported most frequently as the most important reasons not to take full early retirement (Table 2).

Table 1. Working status of the study sample (LASA cohort 2002–2003) at baseline

	% (n)
Have a paid job	n = 999
No	57.9% (578)
Yes	42.1% (421)
Have a (partly) work disability	n = 961[1]
No	77% (740)
Yes	23% (221)
Have taken early retirement	n = 960[1]
No	78.2% (751)
Yes, partly	3.1% (30)
Yes, completely	18.6% (179)
Consider taking early retirement	n = 372[2]
No	36.8% (137)
Yes	63.2% (235)

[1]Due to missing values, the number of respondents is not equal to 999.
[2]This question was asked of those currently working and not having taken early retirement.

Table 2. Frequency of most important reason not to take (full) early retirement

Most important reason reported at baseline	Most important reason not to take early retirement[1]	Most important reason not to take full early retirement[2]
	% (n)	% (n)
Sufficient challenges at work	59.6 (81)	18.1 (40)
Maintain social contacts	17.6 (24)	25.3 (56)
Other pastime less pleasant	0.7 (1)	1.8 (4)
Financially more favourable	5.9 (8)	32.6 (72)
Other reason	16.2 (22)	22.2 (49)
	100% (136)	100% (221)

[1]This question was asked of those with a paid job, who had not taken early retirement, and who were not considering taking early retirement.
[2]This question was asked of those with a paid job, who had not taken early retirement, but who were considering taking early retirement.

Table 3 presents the predictive value of the reasons mentioned for work status at three-year follow-up. It appeared that the majority of those who reported challenges at work as the most important reason not to take (full) early retirement, were indeed still working three years later (84.4% and 66.7%) (Table 3). With respect to the financial advantages as the most important reason not to take full early retirement, it appeared that three years later, 68.3% were indeed still working or partly retired, but a quarter (24.2%) had taken full retirement. The maintenance of social contacts had less predictive value, since one third (35.4%) of those who reported social contacts as the most important reason not to take full retirement, had taken full retirement in the meantime.

Table 3. Working status at follow-up per reason not to take (full) early retirement as reported at baseline

Reason reported at baseline	Working status at follow-up per reason not to take early retirement			Working status at follow-up per reason not to take full early retirement			
	working % (n)	partly/fully retired early % (n)	disabled % (n)	working % (n)	partly retired early % (n)	fully retired early % (n)	disabled % (n)
Enough challenges at work	84.8 (56)	7.6 (5)	7.6 (5)	66.7 (20)	5.9 (2)	11.8 (4)	13.1 (4)
Maintain social contacts	80 (12)	0 (0)	20 (3)	32.5 (13)	10.4 (5)	35.4 (17)	17.5 (7)
Other pastime less pleasant	100 (1)	0 (0)	0 (0)	66.7 (2)	0 (0)	25 (1)	33.3 (1)
Financially more favourable	66.7 (4)	16.6 (1)	16.6 (1)	63.8 (37)	4.5 (3)	24.2 (16)	8.6 (5)
Other reason	84.6 (11)	7.7 (1)	7.7 (1)	46.3 (19)	8.9 (4)	33.3 (15)	9.8 (4)
Total	n = 84	n = 7	n = 10	n = 91	n = 14	n = 53	n = 21

Table 4 describes the most important reasons for taking early retirement among workers as well as among those who had retired early. Among the workers, the pleasure of spending more time on private concerns was by far the most important reason to take early retirement (59.4%). This reason was also reported most frequently by those who had already taken (partial) early retirement (27.3%) (Table 4). Further, among those who had taken early retirement, external factors, such as arrangements that made early retirement attractive and organisational changes, were also reported frequently as the most important reason to have taken early retirement. Health complaints as well as (physical or mental) workload were reported by a only small minority of the workers and retirees (<10%) (Table 4). Although the pleasure of spending more time on private pursuits was reported frequently as the most important reason to take early retirement, the majority (65.7%) were still working at follow-up (data not shown).

Table 4. Frequency table of most important reason to take (full or partial) early retirement

	Workers[1]	Early retirees[2]
Most important reason reported at baseline	% (n)	% (n)
Stress and pressure of work too high	9.8 (23)	6.3 (13)
Physically too heavy	6.8 (16)	5.4 (11)
Health complaints too limiting	6.8 (16)	6.3 (13)
Not motivated anymore	2.1 (5)	5.9 (12)
Nicer to spend more time on private life	59.4 (139)	27.3 (56)
Not possible anymore in the future	1.7 (4)	5.4 (11)
Having worked for many years[3]	2.1 (5)	-
Organisational changes in company[3]	2.6 (6)	9.8 (20)
Arrangements that made early retirement attractive[3]	-	14.6 (30)
Other reason	8.5 (20)	19.0 (39)
	100% (234)	100% (205)

[1] This group included those with a paid job, who had not taken early retirement, but who were considering taking early retirement.
[2] This group included those who have already taken (partial) early retirement.
[3] This category was formed after clustering the answers of "other".

Study 2

Interviews with Workers

Thirty workers were interviewed, either individually or in focus groups. With the exception of the local authority, within each company two interviews were held, each with one worker, as well as one focus group interview with five to eight workers. At the local authority, it was not possible to hold a focus group interview; instead, five workers were interviewed individually. Thus, a total of 11 individual interviews (six men, five women) and three focus group interviews were held among 19 workers (nine men, 10 women) aged 30 to 59 years.

Although the questions in Study 1 differ from those in the qualitative study (i.e. questions related to the reasons either to take or not to take early retirement versus the motivating factors to continue working), the results were rather similar. In line with the LASA results, where only about one third of those currently working were not considering taking early retirement, it appeared from the interviews that most workers were not willing to continue working after the age of 65 years. Although the majority of the interviewees indicated that they were still motivated to work, that they liked their job and that they (still) were healthy enough to perform their job, they did not intend to prolong their working life.

Furthermore, the major reasons (i.e. sufficient challenges at work, mainte-nance of social contacts and the financial aspect) reported by the LASA respon-dents for not taking (full) early retirement were also expressed by the interviewees as motivating factors to continue working. From the responses of the interviews, five key themes were identified: (1) challenges at work, (2) social contacts, (3) re-ward and appreciation, (4) health and (5) competencies and skills (Table 5). The themes include predominantly motivating factors, but also point to measures that can be taken to stimulate a sustained employability.

Table 5. Working status at follow-up per reason not to take (full) early retirement as reported at baseline

Themes	Motivating factors
Challenges at work	- Work climate is important - Being needed, feel oneself useful - Commitment to work and company - Work should be challenging and give satisfaction - Deliver a quality product
Social contacts	- Social contacts - Socially active
Reward and appreciation	- Financial compensation or reward at the sort term - Appreciation for the work done (by giving compliments)
Health	- Prevention of work strain (physically and mentally) - Healthy lifestyle - Optimal balance between work load and capacity
Competencies and skills	- Moving possibilities within company (horizontal and vertical) - Variation in tasks - Career support - Education and training - Coaching role for older worker - Retraining, occupational resettlement

Challenges at Work

Most of the interviewees considered the content of their job of importance to continuing to work. They indicated that they liked their job, were motivated by their work and that they needed their work. With the exception of the workers who performed a physically heavy job, which included routine, it was frequently indicated that they perceived a feeling of satisfaction and motivation when they were being challenged.

By nature, I am rather lazy, but I am challenged by my work. Being at work, I become challenged intellectually; without work, there is no interesting life for me."

"So far, I am not ready to stop. I am motivated to work, to continue work, be-cause the job is challenging enough."

Social Contacts

Without exception, all workers interviewed appeared to set great store on the contact with colleagues and the associated work climate. One worker, for example, expressed this motive as: "Work is both intellectual and social food." Another worker reported: "It can be that your 'world will become so narrow.'.. yes, the contact with colleagues and clients is very important."

Reward and appreciationMost of the interviewees highly valued appreciation from others for the work they did, and considered it as an important factor in continuing. This motive referred to both the financial aspect and reward expressed in words by the supervisor or colleagues. Although none of the interviewees indicated the financial reward as the most important reason to continue working, they agreed that "it definitely plays a role." One worker said: "Respect and appreciation, that's what I think is important."

As to the pat on the back (by the boss) as a motivating factor, they valued receiving a compliment from either the supervisor or colleagues. For example, one worker said:

"I absolutely think reward is essential in remaining motivated to perform the job. This can be through a bonus, but also by your colleagues who say to you how well you performed the task, or by having a dinner together, or something like that."

Health

In the company providing outdoor advertising, the interviewees performed heavy, physical jobs. These workers generally had a negative attitude about prolonged participation in the workforce. One reason for this negative attitude was associated with the total years of having worked when they reached the age of 65 years, since they had started working when quite young. Another reason for their negative attitude concerned the expectation that they would not be able to continue their (current) work, due to the heavy physical workload. Because of their workload, these workers suggested using tools that would reduce the physical work in order to be able to prolong participation in the workforce.

In the remaining three companies, physical workload was not the issue, in contrast to mental workload. Especially in the OHS, workers experienced (too) high work demands. To reduce or cope with work-related stress, some workers suggested implementing a relaxation programme or creating possibilities for relaxation, e.g. by means of a room where workers could rest, or through implementation of a yoga programme.

The promotion of a healthy lifestyle, including physical activity and diet, was also mentioned frequently as being an important factor for increasing the capacity and motivation to prolong a healthy working life. Although they generally agreed that a healthy lifestyle was the worker's own responsibility, they also agreed on the role of the employer in stimulating as well as facilitating such a lifestyle.

"I need to take care that I stay healthy; that's my own responsibility."

Competences and Skills

Finally, the interviewees agreed on the value of education and training of (older) workers to be able to keep up with technological developments. They also reported that training or education was valuable and should be offered by the employer in order to grow (personally), to stimulate challenges at work and to avoid routine work.

"One needs to develop oneself; as soon as the job becomes a routine, it's not good, and one will not remain motivated."

It was further suggested to include the competences and personal development in the functioning discussions:

"In my opinion, the personal development should be included in the yearly functioning discussion."

There were no substantial differences in factors stated by younger and older workers. It appeared only that younger workers had difficulties in describing factors that would motivate them to prolong their working life, as "it is such a long way off."

Study 3

Workshop with OPs and Employers

The OPs and employers generally agreed with the workers' opinions expressed in the interviews. No additional factors were mentioned by them.

As to possible measures to be taken by the employer or the OHS to prolong workers' participation in the workforce, the working groups generally agreed with each other and reported more or less the same measures. From the notes, the following main themes were identified: (1) health promotion, (2) education and training and (3) financial stimuli.

Health Promotion

Each working group independently reported factors that involved promoting the balance between workload and individual capacity, the latter receiving a notable amount of emphasis. The workshop participants not only referred to the promotion of physical activity and exercise, but also emphasised the role of a healthy diet, quitting smoking, a moderate consumption of alcohol and relaxation. Similar to the workers, they agreed on the responsibility of the worker, but also considered the role of the OP and the employer. One working group said, for example:

"It is of importance to stay fit and healthy; this is the responsibility of the worker. The employer, on the other hand, should give the good example. There should be attention for a healthy lifestyle within the organisation."

Another working group expressed its opinion about this issue as follows:

"The employer will do right if he implements a 'vitality policy' including physical activity, fitness, walking in lunchtime or walking during meetings. In most cases, the corporate culture needs to be changed in that it promotes health management with even more stringent measures when neglecting certain activities."

In addition to offering lifestyle programmes and providing information, they considered a periodic health screening to be a useful OHS tool, since the results of such a screening can give direct cause to providing (lifestyle) counselling. With respect to the other side of the balance, i.e. the workload, all agreed that this should be tuned to individual capacity.

Education and Training

The working groups stated that work should be fun and offer sufficient challenges. This could, for example, be achieved by making plans about the work career and education needed and to be followed. Education and training should also be promoted, as it created variation in work, the latter being an important boost to taking pleasure in work. To achieve variation in work, some in the working groups suggested exchanging workers from different companies, or to give older workers a coaching or mentor task in orienting new colleagues.

To illustrate, one working group indicated:

"It is of crucial importance that one enjoys the job! This can be realised by several measures—among others, by giving older workers a coaching task in which they train young workers; the employer can also make agreements with the (older) worker about career planning."

Financial Stimuli

Consistent with the interviews among the workers, attention was paid to the financial aspect. The OPs and employers agreed on the desirability of having both the employer and the government provide financial stimuli to workers who prolonged their working life. Moreover, they advocated maintaining the same net salary when demoting workers because of a (age-related) reduction in work ability.

Discussion

The aim of this study was to examine the reasons for voluntary early retirement as well as for prolonging working life after the official retirement age. Insight into these motives is useful, among other reasons, as input to the HRM policy to retain healthy (older) workers who are able and willing to prolong their labour force participation. Despite the need to tailor the HRM policy to the needs and preferences of older workers [18], it should be kept in mind that older workers are a heterogeneous group in that differences exist in personal characteristics, needs and work ability between individual workers. This was confirmed by the OPs and employers in the present qualitative study, in that they stated that the workload should be tuned to individual capacity.

From the LASA analyses, it was shown that, of those currently working, about two thirds were considering taking early retirement. In view of the economic need to prevent early exit from work, this proportion is substantial. As mentioned before, it is important to encourage workers not to take early retirement, but to prolong working life instead. In order to achieve this, workers should be able as well as be motivated to continue working.

Indeed, the most important reason given by the LASA respondents currently working for not taking early retirement appeared to be the motivation to perform the job, i.e. the presence of challenge at work. This was supported by the interviews among the workers, where pleasure at work was mentioned frequently as a motive to prolong working life. Based on the LASA follow-up data, it appeared that the presence of sufficient challenges at work positively predicted the working status three years later. That is, the majority of workers who found their work challenging, and indicated this was an important reason not to take early retirement, actually remained in the workforce.

Another notable result of this study concerned the fact that the reasons for not taking full early retirement differed from the reasons not to take early retirement (at all). Based on the LASA analyses, the main reasons for the former were the financial aspect of working and the maintenance of social contacts. The social aspects, but particularly the financial advantages of working, thus seemed to play an important role in the decision to either fully retire or to cut down work gradually by partial early retirement.

Again, these results were confirmed by the qualitative study. The large majority of the workers interviewed said the social contacts with colleagues and others at work were of great importance to work motivation. However, based on the LASA follow-up data, it appeared that the maintenance of social contacts had a weak predictive value for the working status three years later. A substantial part of those who had reported social contacts being the most important reason not to take full retirement, did take full retirement in the meantime.

As to the financial aspect, both the workers interviewed and the OPs and employers agreed on its significance. In addition, the financial advantage of not taking retirement appeared to reasonably predict the working status three years later.

Based on these findings, including the desire of the majority of workers to take early retirement, it seems sensible to aim for a gradual exit from work through a period of working fewer hours in order to prevent full early retirement. In this way, the workers benefit financially and moreover can spend some more time on private pursuits, this being the most important reason to take early retirement, as shown by the LASA analyses. This was confirmed by a recent study by Cebulla et al. [19], which stated that there is broad consensus that older workers should be given the opportunity to retire gradually.

Another study found control over the retirement decision to be an important factor in retirement well-being, which persisted three years after retirement [20]. They also found that gradual retirement was followed by a positive change in health. Gradual retirement was linked to an improvement in health. According to the authors of that study, those findings suggest that HR practices that promote employees' control of their retirement decisions will enhance well-being in later life and facilitate prolonged workforce participation [20].

As mentioned in the introduction, not much literature has been published as to the reasons for voluntary early exit from work or for an enduring working life. However, the few previous studies performed in different countries support our results.

For example, Armstrong-Stassen (2006) investigated the importance of six HRM strategies relevant in the retention of older workers [21]. These six factors were: (1) flexible working options, (2) training and development, (3) job design (e.g. reduce workload), (4) recognition and respect, (5) performance evaluation and (6) compensation (e.g. financial, incentives). Although different terms have been used in Table 5, most of the five strategies, if not all, are reflected.

From Armstrong-Stassen's study, it appeared that both retired and employed men rated three HRM practices the same as to their influence on the decision to remain in the workforce: (1) providing challenging and meaningful job tasks; (2) recognising experience, knowledge and skills; and (3) showing appreciation for doing a good job [21]. Further, data from a representative sample of the household population aged 50 and over in England, the English Longitudinal Study of Ageing (ELSA), confirm that influences on retirement are multidimensional, with economic incentives being an important, if not the most important, determinant for continuing working life [22].

Two other factors of influence appeared to be health and social issues. As to the financial aspect as a determinant of retirement behaviour, the literature is consistent [23]. Using ELSA data, Banks et al. (2007) found that both pension accrual and pension wealth are important determinants of the retirement behaviour of men aged 50 to 59. This was also valid among women of the same age, although it was somewhat weaker [24]. It further appeared that there was a U-shaped pattern of being in paid work by quintile of the wealth distribution, with those at the bottom and the top of the wealth distribution being less likely to be in paid work than those in the middle of the wealth distribution, but for different stated reasons: those with relatively low levels of wealth were most likely to stop working due to ill health [24]. Thus, ELSA showed that financial need may act as an incentive to continue working life, but that this varies with wealth, income and education levels [25].

Strengths and Limitations of the Study

As mentioned before, the few previous studies regarding the reasons for early retirement and prolonging working life have mostly used quantitative research methods. One of the strengths of the present study is that it is based on mixed methods, with both quantitative and qualitative data. Such a triangulation method is useful as it approaches the same phenomenon by using different methods, each having its own value. Qualitative research techniques add vast amounts of relevant data [26].

There were, however, some limitations in our qualitative study. Although the transcripts were made by an assistant, the interviews were performed by one researcher, as were the analyses of the responses into codes and themes. Further, despite our not believing that it introduces bias, we did not use a computer software programme in which each item is compared with the rest of the data to establish analytical categories [26]. Instead, the interviewer herself read and reread the recordings, identified the codes and grouped the codes together into key themes. The analyses were performed manually, which was feasible and acceptable, instead of by means of a software package specifically designed for qualitative data management. Although the latter would have made the process easier, we assumed the chance to be small that the manual process would have yielded substantially different clusters. However, as is the case in all qualitative studies, the categorisation of the themes depended on the human factor (i.e. the researcher), and might well have resulted in different clusters if the process had been performed by another researcher.

Another strength of this study is that the quantitative study involved a longitudinal study among a representative sample of older adults in The Netherlands. The LASA follow-up data were useful to provide information about the predictive value of the motives reported at baseline to take (or not) early retirement. Overall, results indicated that the motives reported did not have a high predictive value. However, considering the small number of follow-up data used, prudence is needed for the conclusions as to the predictive value of the reasons mentioned. In this respect, it is also worthwhile to mention that the present study can be considered as a pilot study in that it is based on data of Dutch workers only. As the topic of ageing workers is of worldwide interest, a future study involving diverse countries with diverse policies to encourage a sustained participation in the workforce would be valuable.

Finally, although the involvement of different stakeholders in the qualitative study added value to the study in that extra input could be generated instead of involving only one target group, the number of the representatives of companies was low. Nevertheless, the inclusion of the four companies, which included different in type of workers and work activities, yielded useful information about the organisation of the (HRM) policy with respect to the promotion of an enduring participation in the workforce of older workers.

Conclusion

Taken together, findings suggest that each stakeholder (i.e. the worker, the employer, the OHS and the government) should contribute to the maintenance of a healthy and motivated ageing workforce. In doing so, one should take into account the differences in strategies for different groups, such as different socio-economic status groups or those at different ages. Due to the small size of the study, and especially the qualitative study, the present study could not draw such conclusions for different subgroups. So, further research examining the motivating factors to prolong working life among different groups is recommended.

In addition to their ability to work, workers' motivation is considered to be crucial in sustaining participation in the workforce. Workers who lack pleasure at work and are not motivated to prolong their working life, probably confirm the negative image of some employers about older workers, i.e. lower productivity and "waiting out one's time." There are still some employers for whom the negative effects of ageing prevail over the value of older workers, in that older workers are less productive or resist change or innovations. In order to oppose this view, the employer can and should implement simple measures, such as offering education and expressing appreciation to the personnel. Finally, measures that promote challenges at work, together with financial stimuli, seem to be promising in keeping older workers in the workforce.

Competing Interests

The authors declare that they have no competing interests.

Authors' Contributions

All three authors made a substantial contribution in the design of this specific study. DD was (and still is) involved and responsible for the design and data collection of the quantitative study (Study 1). KP and AvdB were involved in the design and the acquisition of the qualitative data (Study 2). KP performed the statistical analyses. All three authors read and approved the final manuscript for submission to this journal.

References

1. Griffiths A: Ageing, health and productivity: a challenge for the new millennium. Work & Stress 1997, 11:197–214.

2. Van Nimwegen N: Europe at the Crossroads: Demographic Developments in the European Union. Executive Summary, European Observatory on the Social Situation, Demography Monitor 2006 – European Observatory and the Social Situation – Demography Network. Brussels: European Commission; 2006.

3. European Commission: Directorate-General for Employment, Social Affairs and Equal Opportunities Unit E.1. In Europe's Demographic Future: Facts and Figures on Challenges and Opportunities. Luxembourg: Office for Official Publications of the European Communities; 2007.

4. Van Nimwegen N, Beets G: Social Situation Observatory. Demography Monitor 2005. Demographic Trends, Socio-Economic Impacts and Policy

Implications in the European Union. The Hague: Netherlands Interdisciplinary Demographic Institute; 2006.

5. Humphrey A, Costigan P, Pickering K, Stratford N, Barnes M: Factors Affecting the Labour Market Participation of Older Workers. DWP Research Report 2000. Leeds, West Yorkshire: Corporate Document Services; 2003.

6. Meadows P: Retirement Ages in the UK: A Review of the Literature. Employment Relations Research Series No. 18. London: Department of Trade and Industry; 2003.

7. Lund T, Borg V: Work environment and self-rated health as predictors of remaining in work 5 years later among Danish employees 35–59 years of age. Exp Aging Res 1999, 25:429–434.

8. Emmerson C, Tetlow G: Labour market transitions. In Retirement, Health and Relationships of the Older Population in England: The 2004 English Longitudinal Study of Ageing (Wave 2). Edited by: Banks J, Breeze E, Lessof C, Nazroo J. London: Institute for Fiscal Studies; 2006:41–63.

9. Philipson C, Smith A: Extending Working Life: A Review of the Research Literature. DWP Research Report 299. Leeds, West Yorkshore: Corporate Document Services; 2005.

10. Vickerstaff S, Cox J, Keen L: Employers and the management of retirement. Soc Policy Admin 2003, 37:271–287.

11. Krause N, Dasinger LK, Deegan LJ, Brand RJ, Rudolph L: Psychosocial job factors and return to work after compensated low back injury: a disability phase-specific analysis. Am J Industr Med 2001, 40:374–392.

12. Dasinger LK, Krause N, Deegan LJ, Brand JB, Rudolph L: Physical workplace factors and return to work after compensation low back injury: a disability phase-specific analysis. JOEM 2000, 42:323–333.

13. Higgs P, Mein G, Ferrie J, Hyde M, Nazroo J: Pathways to early retirement: agency and structure in the British Civil Service. Ageing Soc 2003, 23:761–778.

14. Parry J, Taylor RF: Orientation, opportunity and autonomy: why people work after state pension age in three areas of England. Ageing Soc 2007, 27:579–598.

15. Deeg DJH, Westendorp-de Serière M: Autonomy and Well-Being in the Aging Population. I. Report from the Longitudinal Aging Study Amsterdam 1992–1993. Amsterdam: Vrije University Press; 1994.

16. Deeg DJH, Van Tilburg T, Smit JH, de Leeuw ED: Attrition in the Longitudinal Aging Study Amsterdam. The effect of differential inclusion in side studies. J Clin Epidemiol 2002, 55:319–328.

17. Britten N: Qualitative research: qualitative interviews in medical research. BMJ 1995, 11:251–253.

18. Agarwal NC: Retirement of older workers: issues and policies. Hum Res Planning 1998, 21:42–51.

19. Cebulla A, Butt S, Lyon N: Working beyond the state pension age in the United Kingdom: the role of working time flexibility and the effects on the home. Ageing Soc 2007, 27:849–867.

20. De Vaus D, Wells Y, Kenid Y, Kendig H, Quine S: Does gradual retirement have better outcomes than abrupt retirement? Results from an Australian panel study. Ageing Soc 2007, 27:667–682.

21. Armstrong-Stassen M: Encouraging retirees to return to the workforce. Hum Res Planning 2006, 29:38–44.

22. Banks J, Casanova M: Work and Retirement. [http://www.ifs.org.uk/elsa/report_wave1.php], In Health, Wealth and Lifestyles of the Older Population in England: The 2002 English Longitudinal Study of Ageing Edited by: Marmot M, Banks J, Blundell R, Lessof C, Nazroo J. London: Institute for Fiscal Studies;

23. Gruber J, Wise D: Social Security Programs and Retirement Around the World: Micro-Estimation. Chicago: The University of Chicago Press; 2004.

24. Banks J, Emmerson C, Tetlow G: Healthy retirement or unhealthy inactivity: how important are financial incentives in explaining retirement? [https://editorialexpress.com/cgi-bin/conference/download.cgi?db_name=res2007&paper_id=427], 2007.

25. Emmerson C, Tetlow G: Labour market transitions. [http://www.ifs.org.uk/publications.php?publication id=3711] In Retirement, Health and Relationships of the Older Population in England: The 2004 English Study of Ageing (Wave 2) Edited by: Banks J, Breeze E, Lessof C, Nazroo J. London: Institute for Fiscal Studies; 2006.

26. Pope C, Ziebland S, Mays N: Qualitative research in health care. Analysing qualitative data. BMJ 2000, 320(7227):114–146.

Worry as a Window into the Lives of People Who Use Injection Drugs: A Factor Analysis Approach

Heidi Exner, Erin K. Gibson, Ryan Stone, Jennifer Lindquist, Laura Cowen and Eric A. Roth

ABSTRACT

Background

The concept of risk dominates the HIV/AIDS literature pertaining to People Who Use Injection Drugs (PWUID). In contrast the associated concept of worry is infrequently applied, even though it can produce important perspectives of PWUID's lives. This study asked a sample (n = 105) of PWUID enrolled in a Victoria, British Columbia needle exchange program to evaluate their degree of worry about fourteen factors they may encounter in their daily lives.

Methods

Exploratory factor analysis was used to analyze their responses.

Results

Factor analysis delineated three factors: 1) overall personal security, 2) injection drug use-specific risks including overdosing and vein collapse and, 3) contracting infectious diseases associated with injection drug use (e.g. HIV/AIDS and hepatitis C).

Conclusion

PWUID in this study not only worry about HIV/AIDS but also about stressful factors in their daily life which have been linked to both increased HIV/AIDS risk behaviour and decreased anti-retroviral treatment adherence. The importance PWUID give to this broad range of worry/concerns emphasizes the need to place HIV/AIDS intervention, education, and treatment programs within a broader harm-reduction framework that incorporates their perspectives on both worry and risk.

Background

Injection drug use is a driving force in historic HIV epidemics in North America and emerging epidemics in Asia and Eastern Europe [1], and is the world-wide leading cause of hepatitis C infection [2]. As a result the public health and epidemiological literature on People Who Use Injection Drugs (PWUID) [3] is dominated by the concept of "risk," associated largely with the sharing of injection drug equipment. Applications of the concept of risk to PWUID originated with seminal studies on disease transmission parameters [4,5], evolved to consider risk networks [6,7], and presently focus on risk environments [8] and the structural production of risk [9].

In contrast to this long-standing concern with the concept of risk there has been relatively little development of a related concept, that of worry, in the HIV/AIDS literature. A notable exception is Smith and Watkins' [10,11] substitution of "worry" about contracting HIV/AIDS in place of "risk of HIV/AIDS." In doing so they argue that worry is an important concept since, "worry is universally experienced and more emotionally based than perceived risk, respondents may have less difficultly understanding the concept of worry and articulating their levels of worry than describing their perceived risk" [10:72]. To this we add that consideration of what people worry about can help identify what individuals and/or groups perceive as actually constituting risks, since risk is now assumed to be socially constructed [9]. From this perspective we can define worry as the recognition of risk. As such consideration of what people worry about has the potential to provide important "windows" into their lives by identifying the risks and challenges they face. This is exemplified by Busza's [12] study of worry patterns for Vietnamese sex workers in Cambodia, which revealed worry about HIV/AIDS nested within broader frameworks encompassing the need to provide funds for extended families in Vietnam, issues of police harassment, and demanding brothel owners.

For PWUID the importance of considering broader factors of concern was highlighted by Mizuno et al. [13] who asked HIV-seropositive PWUID to rank order their "life priorities" from a list of seven items including HIV, housing, having money, food, being able to work, childcare, and safety from violence. In their sample of 161 individuals only 37% ranked HIV as a top priority, while nearly half ranked HIV as their fourth priority or lower. Similarly Brogly et al. [14] asked Montreal PWUID to choose five cards representing what they perceived as the most important factors in their quality of life from a group of seventeen cards including HIV/AIDS treatment, drug treatment, health, being useful, education, feeling good about yourself, independence and free choice, spirituality, friends, family, partnership, sex, housing, money, resources leisure activities, and drugs. From this list the most frequently chosen factors were housing, health, money, spirituality, family, and feeling good about yourself. Only 13% of participants included HIV/AIDS treatment as an important factor, and this factor ranked a lowly twelfth out of the possible seventeen items.

This recognition of worries in addition to HIV/AIDS infection or treatment is important because both ethnographic studies [15] and health surveys [16] report on the chaotic nature of PWUID's lives, with homelessness, stigma, and lack of resources associated with high-risk behaviours ranging from sharing injection drug paraphernalia to survival sex, while simultaneously acting as barriers to HIV/AIDS treatment [17]. In the present paper we analyze responses to a survey questionnaire to delineate worry patterns for PWUID enrolled in a long-established needle exchange program in Victoria, British Columbia in order to gain a broader understanding of what they identify as risk and worry about in their everyday lives.

Methods

Data for this study were generated by a survey of AIDS Vancouver Island's Street Outreach Services (AVI-SOS) Needle Exchange Program clientele. Conducted in April-May/2008 the survey represented a collaborative research project between members of AIDS Vancouver Island and the University of Victoria designed to address the issue of continued injection drug equipment sharing among needle exchange clientele [18]. AIDS Vancouver Island has a long-running Street Out-reach Services (AVI-SOS) Needle Exchange Program. Established in the early 1990s, this service exchanges syringes throughout Vancouver Island. In June 2008 the service was evicted from its fixed Victoria site as a result of a neighbourhood association lawsuit and now operates a mobile needle exchange service in Victoria.

For the present study, eligibility criteria limited participation to persons aged eighteen and over who had injected illicit drugs within the past four months and who were active on the AVI-SOS registry. This registry contains date-specific records of all needle exchanges listed by clients' unique codes. The University of Victoria's Human Research Ethics Board reviewed and approved the survey instrument (Human Subjects' Certificate 08–277). Participants were paid a $20 honorarium for participating in the survey interview, which took less than an hour to administer. The survey included sections pertaining to: 1) basic demographic and educational history, 2) substance use history, 3) current injection practices, 4) egocentric risk networks, 5) worry factors and, 6) drug sharing scenarios.

In total 105 AVI-SOS clientele completed the survey questionnaire. Descriptive statistics for this sample are presented in Tables 1 and 2. These indicate a predominantly male (70%), White (77%), and older (mean age > 40.0 years) sample. Clientele interviewed were characterized by both a long Victorian resi-dency (mean years in Victoria > 17 years) and a lengthy affiliation with the needle exchange (mean time period needle exchange client > 7 years). Despite fairly high levels of formal education and literacy, with almost one-half (48%) completing high school, slightly more than half the interviewees were homeless at the time of the survey (average number of places slept in last week = 2.5) and another 10% were living in shelters. Overall the sample consisted of older, street-entrenched injection drug users with both a long Victoria residence, and a lengthy association with the AVI-SOS needle exchange program.

To investigate worry patterns survey participants scored 14 items on a five-point scale measuring how frequently they worry about each item, with the scale being: 1 = Never, 2 = Once a month, 3 = Weekly, 4 = Daily, 5 = All the time. While response rates to sections within the total survey varied, for example over 10% chose not to complete the network questions, all 105 participants completed this section which was written by AVI-SOS staff in collaboration with needle exchange clientele and pretested to ensure that questions representing risks and challenges Victoria PWUID face daily were included, and that question word-ing and terminology were clear. In the actual questionnaire administration, both individual clients and interviewers had a copy of the questionnaire and went through each section together to ensure mutual understanding of the instrument's questions.

Table 1. AVI-SOS Clientele descriptive statistics for 105 individuals

Variable	N	%
Gender		
Male	74	70
Female	30	29
Transgender	1	1
Ethnicity		
White	81	77
Black	3	3
Hispanic	0	0
Aboriginal	15	14
Metis	6	6
Education		
Grades 1–8	16	15
Attended High School	39	37
Graduated High School	26	25
Post-Secondary	24	23
Housing Situation		
Own/Rent	41	39
Subsidized Housing	10	10
Homeless	54	51

Table 2. Sample size, mean, standard deviation, and range for AVI-SOS clientele descriptive variables

Variable	N	Mean	SD	Range
Age	105	41.6	8.5	19–61
Years lived in Victoria	103	17.3	13.4	0–55
Number of places slept last week	105	2.5	2.0	1–7
Years needle exchange client	105	7.2	5.3	0–19

From discussions with AVI staff and needle exchange clientele prior to administering the questionnaire the 14 items in the worry section were thought to represent three areas of risk and worry for PWUID. These were:

1: Overall Security – this included worry about food and housing as well as personal safety, denoted on the questionnaire as the following items (item number in parentheses): Having a place to stay (8), Able to get food (10), Being robbed (7), Being assaulted (9), Being farmed (robbed while sleeping or high) by your peers (11) and Police arrest (or being jacked up by police) (2).

2: Injection Drug Use-Specific Worries – included here were: Overdosing (1), Vein damage (6), Missing your smash (or vein) (13), Police confiscating syringes (12) and, Getting clean needles (14).

3: Infectious Disease Worries – these include worry about contracting HIV/ AIDS, (3), Hepatitis C Infection (4) and Sexually Transmitted Infections (5).

To assess whether the survey participants also viewed these multiple items in the same perspective we used the SAS° (Version 9.13) PROC FACTOR subroutine to perform an exploratory factor analysis on responses to the worry questions. Factor analysis is a data reduction statistical technique designed to delineate a hypothesized underlying structure of large data sets represented by numerous variables [19]. While our data set cannot be considered large, it does meet an important guideline for factor analysis; that the minimum number of subjects in a sample be either 100 subjects or 5 times the number of variables being analyzed, whichever is larger [20:73]. In our sample 105 participants responded to 14 variables, thus qualifying on both criteria.

Factors are assumed responsible for the covariation between two or more observed variables. Based on either correlation or covariance matrices, factor analysis extracts factors representing shared variance. In this paper, factors were extracted using the maximum likelihood option contained within PROC FACTOR, which permits hypothesis testing for the best number of factors to be retained [20]. Extracted factors were then rotated, that is, a linear transformation was performed on the factor solution for easier interpretation, via the PROMAX option in SAS, rendering the extracted factors correlated or "oblique." Individual variables were considered to "strongly load" on each factor if they possessed factor loading scores equal to or above 0.40 [19:29]. Variables which did not achieve this level (known as low-loading) were removed from subsequent analysis.

To determine the number of factors to be retained in the model a scree test or plot depicting each of the variables as a separate factor with respect to its corresponding eigenvalue (interpreted as the amount of variance accounted for by each factor) was constructed. The point at which the slope of the plot changes from a rapid to a slow decline is the cut-off for the number of factors to be retained. This point separates factors with large eigenvalues from those with relatively small eigenvalues [21]. In addition, as maximum likelihood techniques were used to estimate the factor coefficients, a Chi-square test of the hypothesis that k factors are sufficient was performed.

Results

Univariate Results and Reliability Estimates

To first measure the degree of worry recorded for each variable and to determine if they are associated, we calculated their means and standard deviations. These results are presented in Table 3 with the individual variables placed according to their proposed factor. As seen here there was a wide range exhibited in the mean values for each variable, ranging from the highest value (mean = 3.11) associated with having a place to stay, to the lowest values recorded for being able to obtain clean needles (mean = 1.72) and worry about police arrest (mean = 1.76). Worry about HIV/AIDS had the fourth highest ranking (mean = 2.84) below only worry about having a place to stay, being robbed (mean = 2.91), and contracting hepatitis C (mean = 2.89).

Table 3. Descriptive statistics for individual variables and Cronbach's alpha for each proposed factor

Overall Security Worries[1]		
Variable	Mean	SD
Having a Place to Stay	3.11	1.76
Able to Get Food	2.01	1.41
Being Robbed	2.91	1.63
Being Assaulted	2.51	1.56
Being Farmed	2.68	1.66
Being Arrested	2.76	1.59
Injection Drug Use-Specific Worries[2]		
Variable	Mean	SD
Overdosing	2.04	1.34
Vein Damage	2.65	1.57
Missing Your Smash	2.56	1.70
Police Confiscating Needles	1.76	1.44
Getting Clean Needles	1.72	0.45
Infectious Disease Worries[3]		
Variable	Mean	SD
HIV/AIDS	2.84	1.69
Hepatitis C Infection	2.89	1.71
STIs	2.11	1.50

[1]α = 0.75, [2]α = 0.71, [3]α = 0.74

Also shown in Table 3 is a measure of inter-variable reliability known as Cronbach's alpha, which denotes how well the variables in each proposed factor are related. A commonly applied rule of thumb is that alpha levels should equal or exceed 0.70. As shown in Table 3 this level is met for all three of the proposed factors.

Factor Analysis

Factor analysis proceeded in a two-step manner. First, all fourteen variables were included in the analysis, and the results examined for low-loading variables. Two variables, FARMED and ARREST (Questions 3 and 6 respectively) did not load strongly on any factor. Accordingly, in the second step these variables were removed and the analysis repeated. For this second run the corresponding scree plot was constructed. This revealed a steep drop-off in eigenvalues after the first factor, which had an eigenvalue of 5.14. The second factor featured an eigenvalue of 2.02, while the third factor had an eigenvalue of 0.97. Further, there was not enough evidence to reject the hypothesis that 3 factors are sufficient (X^2_{33} = 33.4, p-value = 0.45). Based on the scree plot and the chi-square test, 3 factors were retained in the model.

Examination of the variable loadings, representing standardized regression coefficients, for each of the three factors is shown in Table 4. These data support our initial supposition of three distinct factors pertaining to: 1) worry about overall personal security, 2) specific worries associated with injection drug use and, 3) worry about contracting infectious diseases associated with injection drug use. The first factor, worry about overall security accounts for over 60% of the total variance and is represented by four manifest variables, being robbed (ROBBED), assaulted (MUGGED), having a place to stay (PLACE) and finding food (FOOD). The second factor contains five variables, worry about overdosing (OD), vein collapse (VEINS), getting clean needles (CLEAN), police taking needles (TAKE) and missing your smash (SMASH), and contributes another 24% of the total variance. The third factor includes the variables relating to worry about contracting HIV/AIDS, hepatitis C (HCV), and sexually transmitted infections (STIs) contributed the remaining 15% of variation.

Table 4. Rotated factor pattern and standardized regression coefficients for the 3 factor model.

Variable	Factor 1	Factor 2	Factor 3
Overdose	0.00	0.41	0.17
HIV	-0.18	0.21	0.67
HCV	0.10	-0.14	0.87
STI	0.09	0.16	0.47
Veins	0.11	0.42	0.28
Robbed	0.70	-0.01	-0.06
Place	0.69	0.01	-0.02
Mugged	0.76	-0.03	-0.08
Food	0.52	0.08	0.03
Take	0.01	0.47	0.08
Smash	0.10	0.74	-0.14
Clean	0.08	0.55	0.06

Coefficients that loaded on a factor (≥0.40) are in bold

Discussion

This paper performed an exploratory factor analysis on data pertaining to a broad array of possible worries thought to characterize the daily life of People Who Use Injection Drugs currently enrolled in a needle exchange program administered by AIDS Vancouver Island, in Victoria, British Columbia, Canada. In doing so it was proposed that three common factors representing worry about overall personal security, health concerns specific to injection drug use, and contracting HIV, HCV, and STIs would be delineated. Exploratory factor analysis indicated that the data did indeed contain these three specific factors. Equally important, each factor fulfilled the four interpretability criteria stressed by Hatcher [19:85–86]: 1) at least three variables with significant loadings on each retained factor, 2) variables loading on a given factor share some conceptual meaning, 3) variables loading on different factors appear to measure different constructs and, 4) the rotated factor pattern demonstrates simple structure, i.e. most variables load on only one factor and have near-zero loadings on others. Two variables, worry about being farmed, or robbed while high by one's peers and worry about police arrest, were dropped from the factor analysis, but univariate analysis showed that they both possessed high mean values, and were viewed as additional important risks to our sample.

Our results are limited in being based on a small non-probabilistic sample which hinders generalization to other settings. However, we note that this analysis corresponds to previous studies [13,14] indicating that for PWUID specific worry about HIV/AIDS exists alongside general living and security considerations.

Consideration of all these concerns echoes the classic paper by Strathdee et al. [22] that argued that "needle exchange is not enough," and that while vital, needle exchange programs should be,"... considered one component of a comprehensive programme including counselling, support and education." More than a decade later these words still ring true, with ethnographic [15] and survey-based [16] studies linking social instability (e.g. homelessness), to both heightened HIV risk behaviour and diminished adherence to anti-retroviral treatment therapy.

While certainly not detracting from the large number of rigorous studies indicating multiple positive HIV/AIDS related harm reduction effects associated with needle exchange programs (for a recent listing of these see [1:143]), our results again emphasize the need to address larger structural problems which form risks and worries for PWUID. Unfortunately in the present case the closure of the AVI-SOS fixed site needle exchange facility limits the organization's ability to address these broader programs. Throughout its existence the fixed-site provided a suite of services, ranging from providing hot meals through access to street nurses, referrals to housing/shelter organizations, personal counsellors and HIV/HCV testing to simply providing a safe, dry, warm place. With its closure AVI-SOS must attempt to provide these services via newly established outreach services which cannot individually offer the array of services provided by the now defunct fixed-site; which this analysis reveals their clientele want and need.

In conclusion, combined with the historically more frequently applied concept of risk, consideration and inclusion of PWUID's panoply of everyday worries into broader-based harm reduction interventions could provide important insights or "windows" into their lives and yield effective programs featuring the convergence of PWUID perspectives and public health goals.

Competing Interests

The authors declare that they have no competing interests.

Authors' Contributions

HE, EKG, JL, RS and EAR designed the study questionnaire, constructed the research design, collected the data upon which this analysis is based, and interpreted analysis results. LC, RS and EAR completed the statistical analysis, and wrote the manuscript draft. All authors read and approved the final manuscript.

Acknowledgements

We wish to particularly thank the participants of this study who gave their time and valuable information. We are also very grateful to AIDS Vancouver Island for their generous use of their facilities for interviewing and overall project support. Financial support was provided by an award from the Vancouver Foundation. EKG is supported by an IMPART Fellowship.

References

1. Institute of Medicine: Preventing HIV Infection among Injecting Drug Users in High-Risk Countries: An Assessment of the Evidence. Washington: National Academies Press; 2007.

2. Aceijas C, Rhodes T: Global estimates of prevalence of HCV infection among injecting drug users. Inter J Drug Policy 2006, 18(5):352–358.

3. Canadian HIV/AIDS Legal Network: "Nothing about us without us." Greater, meaningful involvement of people who use illegal drugs: A public health,

ethical and human rights imperative. [http://www.aidslaw.ca/publications/publicationsdocEN.php?ref=85] 2006.

4. Des Jarlais D, Friedman S, Hopkins W: Risk reduction for the acquired immunodeficiency syndrome among intravenous drug users. Ann Internal Med 1985, 103:775–759.

5. Hagan H, Theile H, Weiss N, Hopkins S, Duchin J, Alexander E: Sharing of drug preparation equipment as a risk factor for Hepatitis C. Amer J Public Health 2001, 91:42–46.

6. Friedman S, Curtis R, Neaigus A, Jose B, Des Jarlais D: Social Networks, Drug Injectors' Lives, and HIV/AIDS. New York, Kluwer Academic; 1999.

7. Neaigus A, Freidman S, Kottiri B, Des Jarlais D: HIV risk networks and HIV transmission among injecting drug users. Evaluation and Program Planning 2001, 24:221–226.

8. Rhodes T, Simpson G, Crofts N, Ball A, Dehne K, Khodakevich I: Drug injecting, rapid HIV spread, and the "risk environment": Implications for assessment and response. AIDS 1999, 13(Supp A):S259–269.

9. Rhodes T, Singer M, Bourgois P, Friedman S, Strathdee S: The structural production of HIV risk among injecting drug users. Soc Sci Med 2005, 61:1026–1044.

10. Smith K: Why are they worried? Concerns about HIV/AIDS in rural Malawi. Demographic Research, Special Collection 2003, 1:277–317.

11. Smith K, Watkins SC: Perceptions of risk and strategies for prevention: Responses to HIV/AIDS in rural Malawi. Soc Sci Med 2005, 60:649–660.

12. Busza J: How does a "risk group" perceive risk? Voices of Vietnam se sex workers in Cambodia. Journal of Psychology and Human Sexuality 2005, 17(1–2):65–82.

13. Mizuno Y, Percell D, Borowski T, Knight K, the SUDIS Team: The life priorities of HIV-seropositive injection drug users: Findings from a community-based sample. AIDS and Behavior 2003, 7(4):395–403.

14. Brogly S, Mercier C, Brunea C, Palepu A, Franco E: Towards more effective public health programming for injection drug users: Development and evaluation of the Injection Drug User Quality of Life Scale. Subst Use and Misuse 2003, 38(7):965–992.

15. Bourgois P: The moral economies of homeless heroin addicts: Confronting ethnography, HIV risk and everyday violence in San Francisco shooting encampments. Subst Use and Misuse 1999, 33:2323–2351.

16. Bouhnik A, Chesney M, Carrieri P, Gallais H, Morneau J, Moatti J-P, Obadia Y, Spire B, the MANIF 2000 Study Group: Nonadherence among HIV-infected injection drug users: The impact of social instability. J Acquired Immun Syndr 2002, 31(Supp 3):S149–S153.

17. Wood E, Kerr T, Tyndall M, Montaner J: A review of barriers and facilitators of HIV treatment among injection drug users. AIDS 2008, 22:1247–1256.

18. Wood E, Tyndall M, Spittal P, Li K, Hogg R, Montaner J, O'Shaughnessey M, Schechter M: Factors associated with persistent high-risk syringe sharing in the presence of an established needle exchange programme. AIDS 2002, 16(6):941–943.

19. Hatcher L: A Step-by-Step Approach to Using SAS for Factor Analysis and Structural Equation Modeling. Cary, NC: SAS Press; 1994.

20. Costello A, Osborne J: Best practices in exploratory factor analysis: Four recommendations for getting the most from your analysis. Practical Assessment, Research Evaluation 2005, 10(7):1–9.

21. Loehlin J: Latent Variable Models: An Introduction to Factor, Path, and Structural Analyses. Hillsdale, NJ: Lawrence Erlbaum Associates, Publishers; 1987.

22. Strathdee S, Patrick D, Currie S, Cornelisse P, Rekart M, Montaner J, Schechter M, O'Shaughnessy M: Needle exchange is not enough: Lessons from the Vancouver injection drug use study. AIDS 1997, 11:F59–65.

Effect of Sunlight Exposure on Cognitive Function Among Depressed and Non-Depressed Participants: A REGARDS Cross-Sectional Study

Shia T. Kent, Leslie A. McClure, William L. Crosson,
Donna K. Arnett, Virginia G. Wadley and Nalini Sathiakumar

ABSTRACT

Background

Possible physiological causes for the effect of sunlight on mood are through the suprachiasmatic nuclei and evidenced by serotonin and melatonin regulation and its associations with depression. Cognitive function involved in these same pathways may potentially be affected by sunlight exposure. We evaluated whether the amount of sunlight exposure (i.e. insolation) affects cognitive function and examined the effect of season on this relationship.

Methods

We obtained insolation data for residential regions of 16,800 participants from a national cohort study of blacks and whites, aged 45+. Cognitive impairment was assessed using a validated six-item screener questionnaire and depression status was assessed using the Center for Epidemiologic Studies Depression Scale. Logistic regression was used to find whether same-day or two-week average sunlight exposure was related to cognitive function and whether this relationship differed by depression status.

Results

Among depressed participants, a dose-response relationship was found between sunlight exposure and cognitive function, with lower levels of sunlight associated with impaired cognitive status (odds ratio = 2.58; 95% CI 1.43–6.69). While both season and sunlight were correlated with cognitive function, a significant relation remained between each of them and cognitive impairment after controlling for their joint effects.

Conclusion

The study found an association between decreased exposure to sunlight and increased probability of cognitive impairment using a novel data source. We are the first to examine the effects of two-week exposure to sunlight on cognition, as well as the first to look at sunlight's effects on cognition in a large cohort study.

Introduction

It is widely accepted that climate and season affect psychological characteristics [1,2]. Recent research has shown that serotonin and melatonin regulation, mechanisms that are involved in the relationship between sunlight and light therapy on mood, are also involved in cognition, which suggests that cognitive function may also be influenced by light [3-5]. Melatonin, serotonin and other mechanisms involved in circadian rhythms are associated with cognitive functioning, and are regulated by the suprachiasmatic nuclei (SCN), which are susceptible to the effects of differing intensities and patterns of environmental illumination [6]. However, the effect of sunlight and light therapy on cognitive function has not been adequately studied. This study aimed to explore if sunlight exposure, measured by insolation (the rate of solar radiation received in an area), is associated with cognitive impairment. In addition, examined the role of season in this relationship.

This study was conducted using baseline data from a large prospective study, the REasons for Geographic And Racial Differences in Stroke (REGARDS) Study, and National Aeronautics and Space Administration (NASA) satellite and ground data. We hypothesized that lower levels of sunlight exposure at participants' residences would be associated with increased rates of cognitive impairment. This study was the first to examine the effects of two-week exposure to natural sunlight on cognition, as well as the first to look at solar effects on cognition in a large cohort study.

Methods

Participants

The present study consisted of participants from the REGARDS study, which has been described in detail elsewhere [7]. In brief, REGARDS is a longitudinal study being conducted to determine relationships between various risk factors and the incidence of stroke. The participants are aged 45 and older and sampled from the 48 conterminous United States. Study participants were oversampled from the "Stroke Belt," a high stroke mortality region consisting of the 8 southeastern states of Arkansas, Louisiana, Tennessee, Mississippi, Alabama, Georgia, North Carolina, and South Carolina. The sample population was particularly oversampled from the "Stroke Buckle," a region with even higher stroke mortality along the coastal plains of Georgia, North Carolina, and South Carolina. Within each region the planned recruitment included half white and half African-Americans (actual: 41% African-American, 59% white). Planned recruitment within each race-region strata was half male and half female (actual total recruitment: 45% male, 55% female). At baseline, a telephone interview was conducted that recorded the patient's medical history, demographic data, socioeconomic status, stroke-free status, depression, and cognitive screening. An in-home exam was conducted recording height, weight, and blood pressure. All participants provided written informed consent, and the study was approved by the Institutional Review Board for Human Subjects at the University of Alabama at Birmingham, as well as all other participating institutions.

Sunlight Exposure (Insolation) Assessment

Sunlight exposure was measured using data values prepared and provided by NASA's Marshall Space Flight Center. Solar radiation values were obtained for 2003 to 2006 from the North American Regional Reanalysis (NARR), an assimilated data product produced by the National Center for Environmental

Prediction (NCEP), a division of the U.S. National Weather Service. The product, including information from satellites and ground observations, was compiled on a grid with a 32 km resolution over North America and matched to each participant by the geocoded residence obtained from the REGARDS database. Solar radiation in Watts/meters^2 (W/m^2), a measure of the instantaneous solar energy reaching the Earth's surface, was assessed 8 times a day at 3-hour intervals for each residence starting at 1:00 AM Pacific Standard Time (PST). A daily integral of solar radiation was calculated; this is referred to as insolation and has units of kilojoules per meters^2 per day (KJ/m^2/day). As a point of reference, under clear skies in late spring or early summer, a typical daily insolation value in the central U.S. is approximately 25,000–30,000 KJ/m^2. In late fall or early winter, a typical daily value is approximately 8,000–10,000 KJ/m^2.

The current residence from the original recruitment file plus updated information from participant at time of scheduling in-home exam was used to geocode each participant. Geocoding of the participants was performed using SAS/GIS batch geocoding. Information obtained from SAS/GIS with 80% accuracy or greater was utilized. Using a subset of the data, we validated the results from the SAS/GIS procedure against a commercially available program http://www.geocode.com using the Haversine formula, and found there to be high agreement between the two algorithms [8]. For those with a SAS/GIS accuracy of 80% or greater, the difference between the latitudes given between the two programs had a mean of 0.23 kilometers and a maximum of 0.95 kilometers.

Cognitive Assessment

A six-item screener questionnaire was used to evaluate global cognitive status by assessing short-term recall and temporal orientation [9-11]. As REGARDS has done in other studies, the score of this screener was dichotomized into an outcome of cognitively impaired or intact. A score of four or fewer correct responses out of the six questions indicated cognitive impairment. Callahan et al. 2002 validated the screener in both a community-based population of 344 black adults aged 65 or older and a population of 651 subjects who were referred to the Alzheimer's disease Center (16% black). Results from the community-based sample found that for a six-item screener score of 4 or fewer, using clinically confirmed cognitive impairment as the gold standard, the sensitivity was 74% and specificity was 80% [10]. The instrument was based on and validated against the Mini-mental State examination [11]. It was also validated against other cognitive measures and diagnoses of both dementia and non-dementia cognitive impairment [10].

Participant Selection

A total of 19,853 participants without previous stroke were enrolled in the study at the time of this analysis (December 1, 2006). Of these, 3,020 patients with poor geocoding (less than 80% accuracy) and a further 33 patients who were missing cognitive scores were excluded, leaving 16,800 participants. Due to data missing in any of the potential confounders, 3,253 of the 16,800 participants were excluded during model selection. Once the model selection was completed, only 2,326 of the 16,800 participants were excluded due to missing covariates selected for the final multivariable model. Chi-squared and t-tests were used to measure differences of the 5,379 excluded and 14,474 included subjects in the final model.

Statistical Analyses

We analyzed insolation from the day the six-item screener was administered and the preceding two weeks. Insolation measurements were analyzed as continuous variables and categorical variables (by 5,000 KJ/m²/day increments).

Due to prior evidence regarding relationships with cognitive function, we considered the following as potential confounders: sex, geographic region (stroke belt, stroke buckle, or non-stroke belt), population density (urban, suburban, and rural), income (less than $20,000, $20,000 to $34,999, $35,000 to $74,900, or $75,000 and more), education (less than high school, high school diploma, some college, or college diploma), race (black or white), smoking (current, past, or never), alcohol use (never used or ever used), Body Mass Index (BMI) (underweight, normal, overweight, or obese), hypertension status (systolic blood pressure ≥ 140, diastolic blood pressure ≥ 90 or self-reported use of hypertension medications), high cholesterol (cholesterol >240), diabetes status (fasting glucose ≥ 126, non-fasting glucose ≥ 200, or self-reported diabetes medications), exercise (weekly or less than weekly), depression status based on the Center for Epidemiologic Studies Depression Scale (CES-D) scale, physical function as measured by the 100 point scale Physical Components Summary (PCS) in the 12-item Short Form (SF-12), season of phone interview (spring, summer, fall, or winter), and age in years (45–54, 55–59, 60–64, 65–69, 70–74, 75–79, or 80 or more) [12,13].

T-tests, chi-squared tests, and correlation tests were used to determine preliminary relationships between insolation, cognition, and the covariates. Cochran-Mantel-Haenzel (CMH) chi-squared tests were used to determine if ordinally categorized insolation had relationships with categorical predictors.

Logistic regressions were used to model the association between insolation and cognition. Backwards elimination was used to build the final multivariable model. Covariates whose relationships with cognitive function carried p-values over 0.10 were not considered for inclusions in the multivariable model. Due to the REGARDS sampling methods, the variables race, region, and sex were included in the multivariable model regardless of statistical significance. All interactions of the remaining covariates with insolation in the final models were considered. For any significant interactions the predicted probabilities and odds ratios (ORs) of cognitive impairment with their 95% confidence intervals were calculated. Since other factors related to seasonality besides insolation may be related to cognitive function, such as temperature, activity level, allergies, and stress, the final model was run both with and without season as a covariate [1,14-17].

Finally, the cognitive screener was divided into two components, the three points that measure short-term recall and the three points that measure temporal orientation. Each of these components was used to explore individual relationships with insolation. Univariate relationships were analyzed using chi-square tests and multivariable relationships were evaluated by taking the final logistic regression model obtained above and replacing the summary measure of cognitive impairment with each of the individual components of the six-item screener. For these analyses, the three point component measures were dichotomized, with one missing point indicating a deficit in either orientation or recall. The measures were also analyzed continuously, so that each component would be equal to the number of points obtained (0, 1, 2, or 3).

Results

Continuous two-week insolation (p = 0.005) but not same-day insolation (p = 0.332) differed significantly by cognitive status (data not shown). Table 1 presents the baseline characteristics overall and by cognitive status. Sex, education, age, income, population density, season, diabetes status, hypertension status, depression status, PCS, alcohol usage, and weekly exercise all differed significantly by cognitive status (all p-values < 0.05; Table 1). CMH chi-squared tests indicated that gender, age, region, population density, season, BMI, PCS-12, and weekly exercise had dose-response relationships with ordinally categorized insolation, but education, diabetes, hypertension, high cholesterol, smoking (data not shown), and depression (Table 2) did not. Dose-response relationships of income (p = 0.0956) and alcohol use (p = 0.0650) with insolation were marginal (data not shown).

Table 1. Demographic, medical, and lifestyle characteristics by cognitive status

Characteristics	All Subjects (N = 16000)	Missing	Intact Cognitive Status (N = 15421; 92%)	Impaired Cognitive Status (N = 1279; 8%)	p-value
	N (%)	N	N (%)	N (%)	
Demographics					
Male	6657 (42)	0	6033 (39)	624 (45)	<0.0001
Education		19			
Less than High School	2081 (12)		1719 (11)	362 (26)	
High School	4404 (26)		3978 (26)	426 (31)	<0.0001
Some College	4525 (27)		4230 (27)	305 (22)	
College Diploma	5761 (34)		5480 (36)	281 (20)	
Age		5			
Less than 55 years	2077 (12)		1982 (13)	95 (7)	
55 to 59 years	2908 (17)		2756 (18)	152 (11)	
60 to 64 years	3197 (19)		3009 (20)	188 (14)	<0.0001
65 to 69 years	3117 (19)		2870 (19)	247 (18)	
70 to 74 years	2418 (14)		2172 (14)	246 (18)	
75 to 79 years	1765 (11)		1536 (10)	229 (17)	
80 or more years	1313 (8)		1091 (7)	222 (16)	
Income		2171			
Less than $20,000/year	3097 (21)		2698 (20)	399 (35)	
$20,000–35,000/year	4017 (27)		3633 (27)	384 (33)	<0.0001
$35,000–$75,000/year	4985 (34)		4670 (35)	298 (26)	
Over $75,000/year	2547 (17)		2467 (18)	80 (7)	
Region		0			
Non Belt/Buckle	7850 (47)		7211 (47)	639 (46)	0.13
Stroke Belt	6095 (36)		5567 (36)	528 (38)	
Stroke Buckle	2855 (17)		2643 (17)	212 (15)	
Population Density		0			
Urban	13532 (81)		12365 (80)	1167 (85)	
Mixed	1678 (10)		1571 (10)	107 (8)	0.0003
Rural	1590 (9)		1485 (10)	105 (8)	
Season		0			
Spring	3610 (22)		3260 (21)	350 (25)	<0.0001
Summer	6439 (38)		5998 (39)	441 (32)	
Fall	3410 (20)		3181 (21)	229 (17)	
Winter	3341 (20)		2982 (19)	359 (26)	
Medical Factors					
BMI		238			
Underweight	212 (1)		189 (1)	23 (2)	
Normal	3973 (24)		3651 (24)	322 (24)	0.26
Overweight	5970 (36)		5462 (36)	508 (38)	

Table 1. *(Continued)*

Obese	6407 (39)		5985 (39)	584 (32)	
Diabetic	3527 (22)	623	3150 (21)	377 (29)	<0.0001
Hypertensive	9854 (59)	73	8943 (58)	911 (67)	<0.0001
High Cholesterol	1740 (10)	80	1595 (10)	145 (11)	0.85
Depressed	1877 (11)	155	1624 (11)	253 (18)	<0.0001
PCS-12 (mean, stddev)	46.1 (10.6)	0	46.3 (10.5)	44.0 (11.0)	<.0001
Lifestyle Factors					
Never Used Alcohol	5129 (31)	0	4606 (30)	523 (40)	<0.0001
No weekly exercise	5948 (36)	234	5359 (35)	589 (44)	<0.0001
Smoking		67			
Current	2445 (15)		2227 (15)	218 (16)	0.18
Past	6609 (40)		6055 (39)	554 (40)	
Never	7679 (46)		7078 (46)	601 (44)	

stddev = standard deviation; BMI = Body Mass Index; PCS = Physical Components Summary
P-values for categorical variables provided from a chi squared test statistic and for continuous variables provided from a scatterthwaite t-test statistic calculated for each variable by cognitive status.
P-values in bold indicate values that are significant at α = 0.05.

Table 2. Crude logistic univariate relationships of depression with two-week insolation

Characteristics	OR (95% CI)
Primary Variable of Interest	
2 week solar radiation (by 5,000 KJ/m²/day)	
<10,000 J/m²	1.14 (0.91–1.42)
10,000–15,000 J/m²	0.94 (0.81–1.09)
15,000–20,000 J/m²	0.87 (0.75–1.01)
20,000–25,000 J/m²	0.87 (0.76–1.00)
>25,000 J/m²	1.00 (Referent)
	CMH chisq p = 0.6054

OR = odds ratio; CI = confidence interval
CMH chisq = Cohran-Mantel-Haenzel chi-squared

Table 3 shows the univariate analyses of categorized insolation testing for dose-response relationships with cognitive function. Two week insolation categorized by 5,000 KJ/m²/day showed does-response effects (p = 0.0075). Participants in the lowest category of insolation compared to those in the highest insolation category had 1.36 times (95% CI = 1.08–1.70) the odds of cognitive impairment. When this measure of insolation was modeled as an ordinal variable, each successively lower insolation level compared to the adjacent higher insolation level had 1.06 (95% CI 1.02–1.11) fold odds of cognitive impairment. This study also confirmed prior study findings that there is a dose-response relationship between cognitive impairment and age, income, and education status (Table 3). Depressed participants showed an increased odds of cognitive impairment (OR = 1.90; 95% CI 1.65–2.20). In addition, univariate analyses showed that the seasons of winter (OR = 1.64; 95% CI 1.42–1.90) and spring (OR = 1.46; 95% CI 1.26–1.69) compared to fall gave increased odds of cognitive impairment.

Table 3. Crude logistic univariate relationships of cognitive impairment with predictors and final covariates

Characteristics	OR (95% CI)
Primary Variables of Interest	
Same day solar radiation (by 5,000 KJ/m²/day)	
<10,000 J/m²	1.15 (0.96–1.38)
10,000–15,000 J/m²	1.02 (0.86–1.21)
15,000–20,000 J/m²	0.96 (0.82–1.13)
20,000–25,000 J/m²	1.03 (0.88–1.21)
>25,000 J/m²	1.00 (Referent)
	CMH chi sq p = 0.4346
2 week solar radiation (by 5,000 KJ/m²/day)	
<10,000 J/m²	1.36 (1.08–1.70)
10,000–15,000 J/m²	1.13 (0.95–1.33)
15,000–20,000 J/m²	1.09 (0.93–1.30)
20,000–25,000 J/m²	1.01 (0.86–1.19)
>25,000 J/m²	1.00 (Referent)
	CMH chi sq p = 0.0075
Covariates	
Male	1.29 (1.15–1.46)
Education	
Less than High School	4.56 (3.57–5.83)
High School	3.36 (2.88–4.17)
Some College	1.91 (1.49–2.46)
College Diploma	1.00 (Referent)

Table 3. *(Continued)*

	CMH chisq p < .0001
Income	
Less than $20,000 per year	4.47 (3.46–5.77)
$20,000 to $35,000 per year	3.31 (2.49–4.15)
$35,000 to $75,000 per year	1.86 (1.45–2.48)
$75,000 or more per year	1.00 (Referent)
	CMH chisq p < .0001
Age	
Less than 55 years	1.00 (Referent)
55 to 59 years	1.15 (0.89–1.50)
60 to 64 years	1.30 (1.01–1.68)
65 to 69 years	1.80 (1.41–2.29)
70 to 74 years	2.36 (1.86–3.02)
75 to 79 years	3.11 (2.43–3.99)
80 or more years	4.25 (3.30–5.46)
	CMH chisq p < .0001
Region	
Non Belt/Buckle	1.00 (Referent)
Stroke Belt	1.07 (0.95–1.21)
Stroke Buckle	0.91 (0.77–1.06)
Season	
Summer	1.00 (Referent)
Fall	0.98 (0.83–1.16)
Winter	1.64 (1.43–1.90)
Spring	1.46 (1.26–1.69)
Depressed	1.90 (1.68–3.26)
PCS-12 (by 10 unit increase)	0.82 (0.78–0.86)
Never Used Alcohol	1.44 (1.28–1.61)

OR = odds ratio; CI = confidence interval; PCS = Physical Components Summary
CMH chisq = Cohran-Mantel-Haenszel chi-squared
ORs in bold have CI's which do not overlap a null value

Due to non-significant (p > 0.10) relationships of high cholesterol, BMI, and smoking with cognitive impairment in crude analyses, these variables were not considered for model-building. Further, population density, diabetes, hypertension, and weekly exercise were dropped from the model for non-significance (p > 0.10) during backwards selection. The final multivariable model included sex, race, education, income, age, region, depression, PCS-12 and alcohol as covariates.

Because depression had a significant interaction with sunlight exposure in this model (p = 0.008), the final model included this interaction term and predicted probabilities of cognitive impairment were calculated according to sunlight exposure category and depression status. Figure 1 shows that the predicted probabilities of cognitive impairment for depressed participants are consistently higher than the predicted probabilities of impairment for non-depressed participants. Figure 1 also shows that depressed participants receiving less than 10,000 KJ/m2/day of sunlight exposure compared to depressed participants in other solar exposure categories had a significantly higher predicted probability of cognitive impairment. When season was added to the multivariable model, it was significantly related to cognitive function (p < 0.01). Both spring (OR = 1.20; 95% CI 1.01–1.42) and winter (OR = 1.33; 95% CI 1.07–1.67) seasons compared to summer season showed increased odds of cognitive impairment. The relationships between sunlight exposure, depression, and cognitive impairment were unchanged when season was added to the model, giving identical predicted probabilities as in Figure 1.

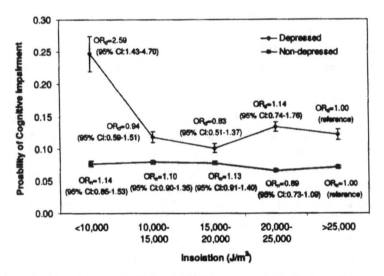

Figure 1. Predicted Probabilities and Odds Ratios of Cognitive Impairment by Depression Status.

Relationships between temporal orientation and short-term recall components of the six-item screener did not reveal any significant univariate or multivariable relationships with insolation, indicating that no single component was likely responsible for the relationship (data not shown).

Conclusion

We found that among participants with depression, low exposure to sunlight was associated with a significantly higher predicted probability of cognitive impairment. This relationship remained significant after adjustment for season. Among participants without depression, insolation did not have a significant effect on cognitive function.

This study adds to the body of literature that shows that environment and lifestyle profoundly affect those who are prone to Seasonal Affective Disorder (SAD) and other types of depression. Studies based on violent homicides, suicides, and aggressive behaviors have repeatedly demonstrated seasonal characteristics, typically with peaks in the spring. These peaks have been associated with sunlight and other climatic variables [18]. Those with SAD have mental states that vary with season, with regular depressions occurring in the winter and remissions in the spring or summer. It is established that these SAD episodes are associated with the shorter daylight hours occurring in winter [19].

The fact that sunlight exposure was associated with cognition in depressed participants supports our hypothesis that the physiological mechanisms which give rise to seasonal depression may also be involved with sunlight's effect on cognitive function. Leonard and Myint, 2006 laid out a paradigm showing how lack of environmental illumination and other stresses might lead to altered serotonin levels, neurodegeneration, depression, cognitive deficits, and ultimately dementia [20]. Both seasonal and non-seasonal depression have been shown to have relationships with environmental illumination [19,21,22]. Theories regarding the body's seasonal cycles, which affect depression and may also affect cognition, are mostly based on the regulation of the body's circadian rhythms by the hypothalamic suprachiasmatic nuclei (SCN) [6,23]. The SCN are modulated by various factors such as body temperature and physical activity, but are in particular modulated by light received by retinal sensors at optimal wavelengths close to sunlight's dominant wavelength of 477 nanometers [23]. The SCN regulate the body's sleep cycle, body temperature, blood pressure, digestion, immune system, and various hormonal systems. Dysfunctional circadian rhythms and sleep disorders, which can occur from inadequate environmental light, have been associated with cognitive deficits [24]. One of the SCN's regulatory functions are their inhibition of the pineal gland from turning serotonin into melatonin during the presence of daytime light [19]. Abnormalities and regulation of both the melatonin and serotonin systems have been found to vary according to sunlight and light therapy in SAD [25,26], bipolar [5] and schizophrenic [27] patients, and even among those without psychiatric diagnoses [28]. Serotonin and melatonin have also been implicated in many mental and cognitive disorders, such as Alzheimer's disease, Parkinson's disease, and sleep disorders [25,29].

Light has been shown to also affect brain blood flow. Cerebral blood flow has specifically been found to improve after phototherapy in pre-term infants [30] and SAD patients [31], and has repeatedly been found to be associated with cognitive functions, such as memory. Inadequate cerebral blood flow has been found to be a likely cause or result of decreasing cognitive functions among those with cardiovascular diseases [32-34], as well as correlated with age-related diseases such as Alzheimer's [35] and non-age related diseases such as Lyme disease [36]. The relationships that serotonin, melatonin, and cerebral hemodynamics have with sunlight, depression, and cognitive function suggest that persons prone to sunlight-related mood disturbances may also be prone to sunlight-related cognitive difficulties.

This study adds to the limited base of knowledge regarding the relationship of weather variables with cognitive function. Studies that have tested the effects of artificial light on cognitive abilities have found that increased light exposure leads to increased alertness and a variety of changes in regional brain activity [37]. In addition, different spectral wavelengths have been found to have differing effects on memory and other cognitive abilities [38]. However, unlike our study, these studies only dealt with immediately acute effects and did not directly examine the effects of natural sunlight. They also have poor generalizability due to using animals or small numbers of human subjects from populations with particular occupations, socio-economic statuses, or ethnicities. We found only two studies that examined the relationship between cognition and sunlight, both of which only dealt with the effects of immediate, short-term exposure. Sinclair et al. (1994) found that increased sunlight exposure was associated with an increase in heuristic processing, which requires memory storage and relevant memory retrieval, but a decrease in systematic processing, a more complicated process requiring analysis and judgment [39]. Keller et al. (2005) found weak positive correlations between sunny days and performance on two measures of cognition, digit span and openness to new information [1]. A major difference between our study and the previous studies is our method of obtaining the participant's exposure to sunlight. The NASA satellite used to obtain the insolation data in this study is able to record data eight times a day as well as provide an accurate characterization of insolation matched to each participant's geocoded home address. This gives superior space and time precision compared to ground sensors used by previous studies. Keller et al. (2005) used barometric pressure as a surrogate for measuring sunny, clear days. Other studies that have not found significant associations between mood or cognition and sunlight in the general population [40] have directly measured insolation using the nearest available ground sensors, which are centered on metropolitan areas. Satellite data allowed us to obtain multiple daily measurements across urban, suburban, and rural areas.

Exposure misclassification exists as a possible limitation of the study. Exposure misclassifications may have taken place if during the two week exposure measurements participants spent a large amount of time in a climate different than the climate recorded by the satellite. This could happen if participants spent large amounts of time indoors or away from their reported home addresses. Also, the daily insolation values were taken by the satellite sensors recorded simultaneously throughout the four different time zones in the U.S. Thus, this point represented different times in the day for different regions of the country. For example, the insolation values used to calculate insolation for participants in the Eastern time zone correspond to 3-hour sampling periods of 1:00, 4:00, 7:00 and 10:00 AM/PM standard time, while for the Mountain time zone the sampling times are 2:00, 5:00, 8:00 and 11:00 AM/PM standard time. However, the relatively short three-hour intervals at which the measurements were taken captures the diurnal cycle well, and the misclassification due to this issue is quite small. It should also be noted that while the relationships found in this study may not apply in younger people (as our study was restricted to those 45 years or older), the participants of the study were recruited from all over the country, with differing demographics, medical factors, and lifestyle factors.

Due to the exclusion of a considerable proportion (27%) of 19,853 enrolled REGARDS participants from the final model as a result of missing values and poor geocoding, we investigated if the excluded participants differed from those with complete information. While sex, education, region, alcohol, age, and depression status of the excluded subjects were statistically different, the proportions of these variables all differed by eight percentage points or less (Table 4). Covariates with larger differences (over 2%) show a disproportionate inclusion into the model of males, those with college diplomas, blacks, non-belt residents, and those that have ever used alcohol. These variables all have known relationships with cognitive impairment and would be the most likely causes of any bias, which might have resulted in underestimating or overestimating the effect of insolation on cognition.

There always remains the possibility of residual confounding. In addition to the imprecision or bias that may be present in any measurement, we could not account for specific psychiatric diagnoses or medicine consumption. Also, environmental temperature may be related to cognitive function, although temperature fluctuations are partially controlled for by season, exercise, cardiovascular factors, and other possible correlates of temperature [41-44]. Eye function is another possible confounder. Specifically crystalline lens transmittance and papillary area have been found affect circadian photoreception, although controlling for age may reduce confounding from these factors. [23]. The interview's time of the day may also have an effect on cognition; however, the sampling method of REGARDS should result in all participants having an equal chance of being interviewed during a given time resulting in similar time distributions at any given variable level [6].

Table 4. Final covariates of excluded and modeled participants

Characteristics	Participants in the Final Model	Excluded Participants	p-value
	N (%)	N (%)	
Total	14,474 (73)	5378 (27)	
Male	5944 (41)	1887 (35)	<.0001
Education			
Less than High School	1670 (12)	798 (15)	
High School	3751 (26)	1525 (28)	<.0001
Some College	3963 (27)	1407 (26)	
College Diploma	5090 (35)	1628 (30)	
Income			
Less than $20,000 per year	3071 (21)	623 (22)	
$20,000 to $35,000 per year	3991 (28)	797 (28)	0.42
$35,000 to $75,000 per year	4904 (34)	914 (32)	
$75,0000 or more per year	2508 (17)	484 (17)	
Age			
Less than 55 years	1885 (13)	585 (11)	
55 to 59 years	2583 (18)	848 (16)	
60 to 64 years	2759 (19)	1052 (20)	<.0001
65 to 69 years	2654 (18)	1038 (19)	
70 to 74 years	2063 (14)	758 (14)	
75 to 79 years	1443 (10)	653 (12)	
80 or more years	1087 (8)	436 (8)	
Black	6291 (44)	2046 (38)	<.0001
Region			

This new finding that weather may not only affect mood, but also cognition, has significant implications and needs to be further elucidated in future studies. That insolation had a relationship with cognitive function but not depression, and that the effect of insolation on cognition is shown among depressed, but not non-depressed participants indicates that insolation may have a relationship with cognition that is independent of, but modified by, depression. It also suggests the possibility that light therapy that is prescribed for SAD may also improve cognitive function. Future studies involving light and other therapies for SAD should include cognitive function as a variable in order to determine relationships with insolation, mood, and cognitive function. Future studies are also needed to demonstrate particular cognitive deficits. The six-item screener was designed to test global cognitive status for large numbers of participants in an easy and efficient manner. While it has adequate sensitivity and specificity as a screening procedure to identify those most likely to have cognitive deficits, it cannot be used to make any particular diagnosis and is limited in its sensitivity to cognitive deficits of small magnitude. In the future, more specific exams and diagnoses can be used to find the specific effects of sunlight on cognitive processes and diseases. We also show that future research regarding treatment and lifestyle should in particular focus on elderly people, since the older a participant is, the more likely the participant is to be cognitively impaired. In addition, research and possibly programs regarding outreach and health education might be targeted to depressives in lower education groups, not only because they are known to have lower access to healthcare in general, but also because they are at a particularly high risk of cognitive impairment. Many of the prior studies have looked at the effects of weather on mood and cognition as seasonal, but the results of this study demonstrate that the effect of season on cognition can be explained by sunlight and other variables. This study also has an interesting finding regarding those without an elevated level of depressive symptoms. We did not find that sunlight meaningfully affected the cognitive abilities of these individuals. However, this lack of a significant finding may be found due to a number of inadequately controlled for indirect behaviors acting as confounders, since there is previous environmental evidence for both season's effects on cognition and environmental illumination's effects on mood and cognition in general populations. Of particular importance, it may be true that those who are non-depressed may spend more time outside, thus receiving a more adequate supply of environmental illumination [17,19,21,22,45-47].

Because cognitive impairment is also associated with other psychological and neurological disorders, discovering the environment's impact on cognitive functioning within the context of these disorders may lead not only to better understanding of the disorders, but also to the development of targeted interventions to enhance everyday functioning and quality of life.

Abbreviations

BMI: Body Mass Index; CESD: Center for Epidemiologic Studies Depression Scale; CMH: Cochran-Mantel-Haenzel; KJ/m2/day: kilojoules per meters² per day; NASA: National Aeronautics and Space Administration; NCAP: National Center for Environmental Prediction (NCEP); NARR: North American Regional Reanalysis; REGARDS: REasons for Geographic And Racial Differences in Stroke; OR: odds ratio; PST: Pacific Standard Time; PCS: Physical Components Summary; SAD: Seasonal Affective Disorder; SF-12: 12-item Short Form; SCN: the suprachiasmatic nuclei; W/m2: Watts/meters².

Competing Interests

The authors declare that they have no competing interests.

Authors' Contributions

SK performed the analysis and drafted the manuscript. LM was a mentor for the methods, statistical analyses, manuscript editing, and data procurement. WC was a consultant for environmental science, manuscript editing, and procured and managed data. DA was a mentor for methods and manuscript editing. VW was a consultant for cognitive function in the REGARDS dataset and manuscript editing. NS was a mentor for the methods, statistical analyses, and manuscript editing.

Authors' Informations

SK is a PhD student in the Department of Epidemiology at the University of Alabama at Birmingham (UAB) and has done work with Marshall Space Flight Center in Huntsville. LM is an Assistant Professor in the Department of Biostatistics at UAB and is also currently working with Marshall Space Flight Center. BC is a scientist working for the Universities Space Research Association and the Marshall Space Flight Center and has extensive experience using satellite data to characterize earth environment variables. VW is an Assistant Professor, works in the Department of Psychology in the School of Medicine at UAB, is the Director of the Dementia Care Research Program, Assistant Director for Translational Research on Aging and Mobility, and has previously used the cohort used in this study for cognitive research. DA is a Professor in and the chair of the Department of Epidemiology at UAB and has extensive experience in cardiovascular and genetic research, an example being the PI of the Genetics of Left Ventricular

Hypertrophy: HyperGEN study. NS is an Associate Professor and an environmental and occupational epidemiologist and pediatrician whose research interests include cancer and infectious diseases epidemiology. Her current research activities include epidemiologic studies relating pesticide exposure and suicide, and of workers in the rubber industry, plastics industry, and chemical manufacturing facilities.

Acknowledgements

This research project is supported by a cooperative agreement U01 NS041588 from the National Institute of Neurological Disorders and Stroke, National Institutes of Health, Department of Health and Human Services. The content is solely the responsibility of the authors and does not necessarily represent the official views of the National Institute of Neurological Disorders and Stroke or the National Institutes of Health. Representatives of the funding agency have been involved in the review of the manuscript but not directly involved in the collection, management, analysis or interpretation of the data The authors acknowledge the participating investigators and institutions for their valuable contributions: The University of Alabama at Birmingham, Birmingham, Alabama (Study PI, Statistical and Data Coordinating Center, Survey Research Unit): George Howard DrPH, Leslie McClure PhD, Virginia Howard PhD, Libby Wagner MA, Virginia Wadley PhD, Rodney Go PhD, Monika Safford MD, Ella Temple PhD, Margaret Stewart MSPH, J. David Rhodes BSN; University of Vermont (Central Laboratory): Mary Cushman MD; Wake Forest University (ECG Reading Center): Ron Prineas MD, PhD; Alabama Neurological Institute (Stroke Validation Center, Medical Monitoring): Camilo Gomez MD, Susana Bowling MD; University of Arkansas for Medical Sciences (Survey Methodology): LeaVonne Pulley PhD; University of Cincinnati (Clinical Neuroepidemiology): Brett Kissela MD, Dawn Kleindorfer MD; Examination Management Services, Incorporated (In-Person Visits): Andra Graham; Medical University of South Carolina (Migration Analysis Center): Daniel Lackland DrPH; Indiana University School of Medicine (Neuropsychology Center): Frederick Unverzagt PhD; National Institute of Neurological Disorders and Stroke, National Institutes of Health (funding agency): Claudia Moy PhD.

Additional funding, data, data processing, and consultation were provided by an investigator-initiated grant-in-aid from NASA. NASA did not have any role in the design and conduct of the study, the collection, management, analysis, and interpretation of the data or the preparation or approval of the manuscript. The manuscript was sent to NASA Marshall Space Flight Center for review prior to submission for publication.

References

1. Keller MC, Fredrickson BL, Ybarra O, Cote S, Johnson K, Mikels J, Conway A, Wager T: A warm heart and a clear head. The contingent effects of weather on mood and cognition. Psychol Sci 2005, 16:724–731.

2. Sinclair RC, Mark MM, Clore GL: Mood-related persuasion depends on (mis) attributions. Social Cognition 1994, 12:309–326.

3. Winkler D, Pjrek E, Iwaki R, Kasper S: Treatment of seasonal affective disorder. Expert Rev Neurother 2006, 6:1039–1048.

4. McColl SL, Veitch JA: Full-spectrum fluorescent lighting: a review of its effects on physiology and health. Psychol Med 2001, 31:949–964.

5. Srinivasan V, Smits M, Spence W, Lowe AD, Kayumov L, Pandi-Perumal SR, Parry B, Cardinali DP: Melatonin in mood disorders. World J Biol Psychiatry 2006, 7:138–151.

6. Van Someren EJ, Lek RF: Live to the rhythm, slave to the rhythm. Sleep Med Rev 2007, 11:465–484.

7. Howard VJ, Cushman M, Pulley L, Gomez CR, Go RC, Prineas RJ, Graham A, Moy CS, Howard G: The reasons for geographic and racial differences in stroke study: objectives and design. Neuroepidemiology 2005, 25:135–143.

8. Sinnott RW: Virtues of the Haversine. 1984, 68:159.

9. Wadley VG, McClure LA, Howard VJ, Unverzagt FW, Go RC, Moy CS, Crowther MR, Gomez CR, Howard G: Cognitive status, stroke symptom reports, and modifiable risk factors among individuals with no diagnosis of stroke or transient ischemic attack in the REasons for Geographic and Racial Differences in Stroke (REGARDS) Study. Stroke 2007, 38:1143–1147.

10. Callahan CM, Unverzagt FW, Hui SL, Perkins AJ, Hendrie HC: Six-item screener to identify cognitive impairment among potential subjects for clinical research. Med Care 2002, 40:771–781.

11. Folstein MF, Folstein SE, McHugh PR: "Mini-mental state." A practical method for grading the cognitive state of patients for the clinician. J Psychiatr Res 1975, 12:189–198.

12. Ware J Jr, Kosinski M, Keller SD: A 12-Item Short-Form Health Survey: construction of scales and preliminary tests of reliability and validity. Med Care 1996, 34:220–233.

13. Melchiot LA, Huba GJ, Brown VB, Reback CJ: A short depression index for women. Educational and Psychological Measurement 1993, 53:1117–1125.

14. Palinkas LA, Makinen TM, Paakkonen T, Rintamaki H, Leppaluoto J, Hassi J: Influence of seasonally adjusted exposure to cold and darkness on cognitive performance in circumpolar residents. Scand J Psychol 2005, 46:239–246.

15. Blaiss MS: Cognitive, social, and economic costs of allergic rhinitis. Allergy Asthma Proc 2000, 21:7–13.

16. Marshall PS, Colon EA: Effects of allergy season on mood and cognitive function. Ann Allergy 1993, 71:251–258.

17. Paakkonen T, Leppaluoto J, Makinen TM, Rintamaki H, Ruokonen A, Hassi J, Palinkas LA: Seasonal levels of melatonin, thyroid hormones, mood, and cognition near the Arctic Circle. Aviat Space Environ Med 2008, 79:695–699.

18. Lambert G, Reid C, Kaye D, Jennings G, Esler M: Increased suicide rate in the middle-aged and its association with hours of sunlight. Am J Psychiatry 2003, 160:793–795.

19. Miller AL: Epidemiology, etiology, and natural treatment of seasonal affective disorder. Altern Med Rev 2005, 10:5–13.

20. Leonard BE, Myint A: Changes in the immune system in depression and dementia: causal or coincidental effects? Dialogues Clin Neurosci 2006, 8:163–174.

21. Espiritu RC, Kripke DF, Ancoli-Israel S, Mowen MA, Mason WJ, Fell RL, Klauber MR, Kaplan OJ: Low illumination experienced by San Diego adults: association with atypical depressive symptoms. Biol Psychiatry 1994, 35:403–407.

22. Haynes PL, Ancoli-Israel S, McQuaid J: Illuminating the impact of habitual behaviors in depression. Chronobiol Int 2005, 22:279–297.

23. Turner PL, Mainster MA: Circadian photoreception: ageing and the eye's important role in systemic health. Br J Ophthalmol 2008, 92:1439–1444.

24. Walker MP, Stickgold R: Sleep-dependent learning and memory consolidation. Neuron 2004, 44:121–133.

25. Khait VD, Huang YY, Malone KM, Oquendo M, Brodsky B, Sher L, Mann JJ: Is there circannual variation of human platelet 5-HT(2A) binding in depression? J Affect Disord 2002, 71:249–258.

26. Leppamaki S, Partonen T, Vakkuri O, Lonnqvist J, Partinen M, Laudon M: Effect of controlled-release melatonin on sleep quality, mood, and quality of life in subjects with seasonal or weather-associated changes in mood and behaviour. Eur Neuropsychopharmacol 2003, 13:137–145.

27. Jakovljevic M, Muck-Seler D, Pivac N, Ljubicic D, Bujas M, Dodig G: Seasonal influence on platelet 5-HT levels in patients with recurrent major depression and schizophrenia. Biol Psychiatry 1997, 41:1028–1034.

28. Golden RN, Gaynes BN, Ekstrom RD, Hamer RM, Jacobsen FM, Suppes T, Wisner KL, Nemeroff CB: The efficacy of light therapy in the treatment of mood disorders: a review and meta-analysis of the evidence. Am J Psychiatry 2005, 162:656–662.

29. Srinivasan V, Pandi-Perumal SR, Cardinali DP, Poeggeler B, Hardeland R: Melatonin in Alzheimer's disease and other neurodegenerative disorders. Behav Brain Funct 2006, 2:15.

30. Dani C, Bertini G, Martelli E, Pezzati M, Filippi L, Prussi C, Tronchin M, Rubaltelli FF: Effects of phototherapy on cerebral haemodynamics in preterm infants: is fibre-optic different from conventional phototherapy? Dev Med Child Neurol 2004, 46:114–118.

31. Matthew E, Vasile RG, Sachs G, Anderson J, Lafer B, Hill T: Regional cerebral blood flow changes after light therapy in seasonal affective disorder. Nucl Med Commun 1996, 17:475–479.

32. Jennings JR, Muldoon MF, Price J, Christie IC, Meltzer CC: Cerebrovascular support for cognitive processing in hypertensive patients is altered by blood pressure treatment. Hypertension 2008, 52:65–71.

33. Beason-Held LL, Moghekar A, Zonderman AB, Kraut MA, Resnick SM: Longitudinal changes in cerebral blood flow in the older hypertensive brain. Stroke 2007, 38:1766–1773.

34. Elias MF, Sullivan LM, Elias PK, D'Agostino RB Sr, Wolf PA, Seshadri S, Au R, Benjamin EJ, Vasan RS: Left ventricular mass, blood pressure, and lowered cognitive performance in the Framingham offspring. Hypertension 2007, 49:439–445.

35. Scarmeas N, Zarahn E, Anderson KE, Habeck CG, Hilton J, Flynn J, Marder KS, Bell KL, Sackeim HA, Van Heertum RL, et al.: Association of life activities with cerebral blood flow in Alzheimer disease: implications for the cognitive reserve hypothesis. Arch Neurol 2003, 60:359–365.

36. Fallon BA, Keilp J, Prohovnik I, Heertum RV, Mann JJ: Regional cerebral blood flow and cognitive deficits in chronic lyme disease. J Neuropsychiatry Clin Neurosci 2003, 15:326–332.

37. Vandewalle G, Balteau E, Phillips C, Degueldre C, Moreau V, Sterpenich V, Albouy G, Darsaud A, Desseilles M, Dang-Vu TT, et al.: Daytime light exposure dynamically enhances brain responses. Curr Biol 2006, 16:1616–1621.

38. Vandewalle G, Gais S, Schabus M, Balteau E, Carrier J, Darsaud A, Sterpenich V, Albouy G, Dijk DJ, Maquet P: Wavelength-dependent modulation of brain responses to a working memory task by daytime light exposure. Cereb Cortex 2007, 17:2788–2795.

39. Chen S, Duckworth K, Chaiken S: Motivated Heuristic and Systematic Processing. Psychological Inquiry 1999, 10:44–49.

40. Bulbena A, Pailhez G, Acena R, Cunillera J, Rius A, Garcia-Ribera C, Gutierrez J, Rojo C: Panic anxiety, under the weather? Int J Biometeorol 2005, 49:238–243.

41. Hansen A, Bi P, Nitschke M, Ryan P, Pisaniello D, Tucker G: The effect of heat waves on mental health in a temperate Australian city. Environ Health Perspect 2008, 116:1369–1375.

42. Makinen TM, Palinkas LA, Reeves DL, Paakkonen T, Rintamaki H, Leppaluoto J, Hassi J: Effect of repeated exposures to cold on cognitive performance in humans. Physiol Behav 2006, 87:166–176.

43. Huang J: Prediction of air temperature for thermal comfort of people in outdoor environments. Int J Biometeorol 2007, 51:375–382.

44. Alperovitch A, Lacombe JM, Hanon O, Dartigues JF, Ritchie K, Ducimetiere P, Tzourio C: Relationship between blood pressure and outdoor temperature in a large sample of elderly individuals: the Three-City study. Arch Intern Med 2009, 169:75–80.

45. Booker JM, Roseman C: A seasonal pattern of hospital medication errors in Alaska. Psychiatry Res 1995, 57:251–257.

46. Brennen T, Martinussen M, Hansen BO, Hjemdal O: Arctic cognition: a study of cognitive performance in summer and winter at 69 degrees N. Appl Cogn Psychol 1999, 13:561–580.

47. Mills PR, Tomkins SC, Schlangen LJ: The effect of high correlated colour temperature office lighting on employee wellbeing and work performance. J Circadian Rhythms 2007, 5:2.

Safe Using Messages may not be Enough to Promote Behaviour Change Amongst Injecting Drug Users Who are Ambivalent or Indifferent Towards Death

Peter G. Miller

ABSTRACT

Background

Health promotion strategies ultimately rely on people perceiving the consequences of their behaviour as negative. If someone is indifferent towards death, it would logically follow that health promotion messages such as safe using messages would have little resonance. This study aimed to investigate attitudes towards death in a group of injecting drug users (IDUs) and how such attitudes may impact upon the efficacy/relevance of 'safe using' (health promotion) messages.

Methods

Qualitative, semi-structured interviews in Geelong, Australia with 60 regular heroin users recruited primarily from needle and syringe programs.

Results

Over half of the interviewees reported having previously overdosed and 35% reported not engaging in any overdose prevention practices. 13% had never been tested for either HIV or hepatitis C. Just under half reported needle sharing of some description and almost all (97%) reported previously sharing other injecting equipment. Many interviewees reported being indifferent towards death. Common themes included; indifference towards life, death as an occupational hazard of drug use and death as a welcome relief.

Conclusion

Most of the interviewees in this study were indifferent towards heroin-related death. Whilst interviewees were well aware of the possible consequences of their actions, these consequences were not seen as important as achieving their desired state of mind. Safe using messages are an important part of reducing drug-related harm, but people working with IDUs must consider the context in which risk behaviours occur and efforts to reduce said behaviours must include attempts to reduce environmental risk factors at the same time.

Background

Injecting drug users (IDUs) experience higher rates of death and poorer health than their non-injecting peers. IDUs are between 6 and 20 times more likely to die than their non-heroin-using peers of the same age and gender [1]. Death due to suicide among heroin users occurs at 14 times the rate of matched peers [2]. The major type of heroin-related mortality and morbidity is heroin-related overdose. At the time of this study, the number of deaths attributed to opioid overdose in Victoria had risen from 49 in 1991 to 331 in 2000. In Australia, around a quarter of heroin users report having experienced an overdose in the past 6 months, and over 70% reporting having witnessed an overdose in the previous 12 months [3-5]. The other major cause of mortality and morbidity in IDUs is the transmission of blood-borne viruses (BBVs), most usually HIV and hepatitis C (HCV). The high prevalence of HCV infection, and the increased infective ability of HCV in comparison to HIV, makes sharing of all forms of drug paraphernalia, not simply needles, a high-risk practice [6]. In addition to the risk of overdose and BBV transmission, environmental factors such as the illicit status of heroin, stigmatisation of IDUs and barriers to effective treatment maximise the consequences of risky behaviour. These factors combine to create an environment where death and disability are common occurrences for IDUs and this study seeks to document IDU attitudes towards death and the relationship between these attitudes and health promotion strategies.

Health promotion strategies (such as health education programs) have shown some success in the general population and much of this thinking has influenced the programs implemented with IDUs such as 'safe using messages' aimed at preventing overdose and BBV transmission. However, there is a small, but growing, literature which documents examples of when human desires and preferences mean that health behaviour is prioritised lower than other considerations. This has been seen in regard to the use of condoms [7,8], dietary habits [9,10] and smoking [11,12]. The majority of interventions targeted at overdose have revolved around 'safe using messages.' Typical messages include: 'don't mix your drugs,' 'split the dose,' 'always use with a friend,' 'use where you can be found' and 'watch your tolerance' [13]. Whilst there is abundant literature describing program implementation of safe using messages, there are no evaluative studies of such strategies. The main intervention targeted at reducing BBV transmission has been needle and syringe programs (NSPs) and their associated safe using messages. Such messages include: 'don't share needles,' 'don't reuse needles' and 'don't share other injecting equipment.' Because of the logically combined nature of these interventions, the effectiveness of health promotion messages alone remains untested, but such programs appear to have limited success in reducing some harms compared to others, especially in relation to overdose prevention and HCV transmission. While there have been some investigations around risk behaviour in marginalised groups [e.g. [14,15]], these studies have not investigated the role of attitudes towards death and how indifferent attitudes affect the relevance of health promotion messages. This study sought to understand some potential barriers for IDUs acting on health information, investigating their attitudes towards drug taking and death and how such attitudes may impact upon the effectiveness of safe using messages.

IDUs' Attitudes Towards Risk

Most IDUs report never or rarely worrying about overdose or BBV transmission (excluding HIV/AIDS) [14,16]. Though not well studied, prior studies have also shown that engaging in high risk behaviours does not necessarily mean that someone has a reduced fear of death [17]. In some instances, individuals will act to minimise risk that is an unavoidable part of their environment, while still engaging in risky behaviours. For instance, crashes and death in serious recreational cyclists, a pursuit that involves regular brushes with death, are viewed as inevitable or unavoidable and are seen as 'occupational hazards' [18]. Although the high level of danger is constant, cyclists are not actually 'death cheaters.' Rather, "due to the unavoidably risk-laden nature of the activity, the subculture of cycling has incorporated the dangers of riding in ways that inextricably linked them to the very enactment of that life, the bike life" [[18]: 169]. Many risk takers, (e.g. parachutists and cyclists) often carefully try to reduce the risk as far as possible, but in some cases, such as cyclists, environmental factors such as the dominance of cars on the road, mean that the hazards they are exposed to are substantially increased and beyond their control. Similar attitudes have been theorised for soldiers in conflict situations, particularly those from lower class backgrounds [19,20]. In their case, it has been suggested that indifference towards death is socially constructed through the dual masculinised roles of both "a man" who carries arms, trained to kill and to cope with the death of a close friend, or a "real man" who takes care of, and provides for, his family [19]. Both roles ultimately view death as an occupational hazard, though, like cyclists, they are not indifferent to their fates and take all reasonable precautions.

The Social Risk Environment

The perception of risk is highly contextual and it is worth considering that risk can not only be enjoyed or avoided, it can also be ignored. For example, Plumridge and Chetwynd [21] also observed that, for some of their sample, risk was not denied or overridden, but acknowledged. This can also be affected by the individual's self constructed identity and the social environment they inhabit. They noted that IDUs can inhabit "a social world in which there was very little sense in which anything other than drug taking provided a raison d'etre" [21].

People are often driven by ambivalent and confused motives, such as a desire to achieve relief from pain or to escape an unbearable situation [15,21,22]. While most individuals will reject the role of social/structural determinants on their behaviour, preferring individualistic explanations that affirm self-efficacy [21,23], research consistently identifies how the social and structural environment we inhabit influences our behaviour, particularly in relation to drugs [23,24]. Specifically, the relationship between poverty, its consequent marginalisation and risky drug-taking behaviour is well documented [25,26].

This relationship is even stronger when urban deprivation is found in combination with vulnerability and trauma [27,28]. Deprived urban settings are often violent and depressed contexts in which hope of attaining socially ordained norms such as career, wealth and status are only attained by the token few. In such settings, risk and death can be less unattractive than a desire to relieve existential pain, or escape a sense of hopelessness [29]. Importantly, such drug use and attitudes towards risk reflect the reality that drug use can be functional, pleasurable, problematic and dangerous at the same time [26]. Within such a personal and social milieu, reduction of harm may not be prioritised.

Methods

Sixty heroin users were interviewed over a six week period in April/May 2000 at two needle and syringe programme (NSP) sites in Geelong, Australia. The sample was a convenience sample and interview subjects were recruited using contact cards handed out by outreach workers, NSP workers and ambulance paramedics attending overdose events. The recruitment card informed potential participants that interviews were about risk and heroin use. To be eligible for the study, subjects had to have used heroin in the previous month. Interviews were conducted in interview rooms provided by Barwon Health Drug and Alcohol Services. Ethical clearance was granted by both Deakin University Human Research Ethics Committee and Barwon Health Research Ethics Committee. Access to counselling was provided if required as well as referral for other support services. No interviewees requested counselling, although one interviewee was referred to the local psychiatric service following a suicide attempt. Subjects were reimbursed $20 per interview.

Qualitative, semi-structured interviews were used and interviewees were encouraged to talk freely of their experiences and opinions. General discussion topics of interest were listed on a checklist to ensure all interviewees views were sought on each topic. Discussions were not structured in any particular order and topics were ticked off as mentioned in the normal course of the more general discussion. All interviews were recorded and transcribed verbatim. Interviews took between 20 and 95 minutes and subjects were required to use a pseudonym to ensure anonymity. Participants were asked about overdose patterns, blood-borne virus behaviour, suicidality and attitudes towards death [29-31]. They were specifically asked about their risk behaviours, attitudes toward death, whether they had ever attempted suicide and were engaged in subsequent conversation regarding details on each topic such as triggering events and other contextual details. While the study also looked at suicidal thoughts and behaviour, these findings are presented elsewhere [29]. All questions were read out during the interview.

Setting

Geelong is a city of approximately 205,000 people with a growth rate of 1.1% per annum. Located 70 kilometres from Melbourne, it is both a regional centre and a suburb of Melbourne. Geelong is traditionally and industrial and port town, but has seen massive decline since the 1970s and now has few large manufacturers remaining. This working class basis and subsequent decline in employment has seen a raft of social problems over the past 3 decades, with alcohol, drugs and drug-related violence featuring prominently on the social landscape. A number of traditional working-class suburbs have become dominated by social housing, unemployment and social security dependence. Most interviewees reported currently living in these suburbs, although it is unclear how long they have lived there and over half reported unstable housing.

Analysis

Statistical analysis was conducted with SPSS and qualitative data was analysed using NVivo. The narratives in this article result from thematic categorisation. Thematic analysis is an inductive design where, rather than approach a problem with a theory already in place, the researcher identifies and explores themes which arise during analysis of the data [32]. In this analysis, once a theme became evident, all transcripts were reanalysed for appearances of the theme. Categorisation was not exclusive and some narratives appeared in many themes. Categories are added to reflect as many of the nuances in the data as possible, rather than reducing the data to a few numerical codes [33]. All the data relevant to each category were identified and examined using a process called constant comparison, in which each item is checked or compared with the rest of the data to establish analytical categories. For the sake of transparency, results reported are enumerated [34]. Where available, narratives which present opposing viewpoints will also be presented [35].

Limitations

The aim of this article is not to present an exhaustive analysis of this data, but to offer some insights on indifference and injecting drug use using this qualitative material. To this end, and considering the relatively small sample size, the findings presented here are not generalisable. In addition to this, the thematic coding undertaken was conducted by a single researcher and may therefore be open to interpretation. The study is also limited in terms of its limited geographical range and the possibility that different localities will carry different cultures around risk, although this was not evident in the available comparisons such as rates of over-dose and needle-sharing. The study also lacked a stated sampling frame, simply using a convenience sample of people who attended NSPs. It is possible that more a more structured sampling frame, combined with a larger sample, may have identified differences within sub-groups of IDUs in relation to risk behaviour and attitudes towards death.

Finally, the study ultimately relied on self report. While self report has been found reliable in relation to behaviours which are able to be measured though other means [36], it is unwise to assume that self report will be reli-able for all aspects of a person's behaviour. In particular, when talking about death and risk, it is possible that a number of interviewees displayed some bravado or other reasons for reporting in a socially constructed manner. Previ-ous research has identified that there are many factors which might affect the way in which interviewees wish to present themselves. Interviews are firstly a socially interactive enterprise. "Evidence of such reflexive organisation of the self can be seen in individuals' sensitivity to social circumstance and sanction in relation to their identities" [21]. Motivations can include the preserva-tion of personal self constructions such as heroic individualism, responsibility, maturity, courage or weakness which ultimately reflect their sense of moral worth. Ultimately, interview accounts are constructed by actors interested in achieving certain social effects in their story-making concerning identity, reflexive biography and, for the purpose of this study, agency concerning risk management and attitudes towards death [37,38]. On the other hand, it is worth considering that most self report data has aligned with other research evidence [21].

Results

Most of the interviewees (n = 36) were male (see Table 1). The average age of interviewees was 28.1 years old (range 15–51 years). All interviewees had used heroin within the past week and most reported that their main 'drug of choice' was heroin. Over half (53%) of interviewees were not currently in treatment and 30% were in methadone maintenance treatment (MMT).

Overdose Experiences and Prevention

Over half of the 60 interviewees (n = 35, 58%) report having previously over-dosed, with an average of 4 (SD = 3.79) previous overdoses. Thirty two percent (n = 19) of interviewees reported doing nothing to prevent overdose.

Table 1. Summary Statistics

Mean Age (range) yrs	28.1 (15–51)
Median Age yrs	26.0
N Male (%)	36 (60%)
Education, N (%)	
- year 10 or less N (%)	33 (55%)
- commenced university	3 (5%)
Employment, N (%)	
-unemployed	36 (60%)
-pension/disability support	12 (20%)
-part-time employed	9 (15%)
-students	3 (5%)
Accommodation, N (%)	
-homeless	22 (36%)
-rental	19 (32%)
Drug of choice, N (%)	
- Heroin	55 (92%)
- Amphetamines	3 (5%)
- Cannabis	2 (3%)
Heroin Use Duration	
-Mean (SD)	7.4 yrs (7.37)
-Range	1–30 years
Treatment, N (%)	
-not in treatment	32 (54%)
-methadone maintenance	18 (30%)
-counselling	6 (10%)
Overdose:	
- At least once	35 (58%)
- mean (SD)	4 (SD = 3.79)
-median	3
-range	1–15
Blood borne viruses:	
- HIV tested	54 (90%)
-HCV tested	52 (87%)
- HIV +ve	0
- HCV +ve	32 (54%)
- HBV +ve	2 (3%)

BBV Rxperience and Risk Behaviours

Interviewee behaviour regarding testing and risk behaviour around BBVs can be seen as possible indicators of the behaviours are able to engage in if they are not ambivalent towards their own fate, as well as a measure of the harm they have already experienced. A substantial proportion of interviewees were unaware of their HIV or HCV serostatus (13% and 10% respectively). Over half (54%, n = 32) were HCV positive and none were HIV positive. Around one in five of the interviewees in this study self-reported both ever borrowing someone else's needle (18%) and lending their needle to someone else (22%).

Attitudes Towards Death

Two questions about death were asked. The first question asked the participant whether or not they ever talked about death with their peers. Most (84%, n = 50) reported that they never talked about death, although 3% (n = 2) reported that they often discussed death as a possible consequence of their heroin use. Interviewees were also asked how they felt about death and whether they were afraid of dying. The vast majority (82%, n = 49) stated that they were never afraid of dying, 12% (n = 7) said that they were afraid of dying from some causes other than heroin use (i.e. car accident) and 3% (n = 2) of the interviewees reported that they were often afraid of dying. Narrative responses showed that almost half of the interviewees (n = 28) were either indifferent or fatalistic about death.

Wayne, 51 yrs, Well, I surely don't want to die, but it doesn't make me not want to use. If it did I wouldn't use any more, because I've dropped a few times. It hasn't frightened me off enough. I know if I die, I'll just go to sleep any way, I just don't wake up.

Wayne's narrative provides an example where overdose death is perceived to be a comparatively pleasant experience. This attitude can be seen in its extreme form in the following narrative.

Casey, 15 yrs, I reckon that was the best feeling, overdosing. The best feeling ever. The first time I ever felt so stoned. It was just the best feeling ever. There was a time when I was apparently dead. It was grouse, I felt like a was asleep and I was just going through this full trippyness. It was the best feeling.

Casey's narrative holds a number of insights into both the motivation for risky heroin use, but also could be an example of the bravado expressed by a young person discussing a frightening experience. In the context of a research interview, and the complexities of such a social interaction, it is probable that both elements are at play. Ten interviewees also reported indifference towards both life and death.

Peter, 28 yrs, ... sometimes it gets too much. You're broke all the time. You haven't got a roof over your head or you haven't got money for food. You just get sick of the lifestyle. It's a real bugger because it's something you love but you get discriminated against. You know, the way people treat you, even your family. It [heroin overdose] would be a good way to go, better than cancer.

Peter's narrative points to many factors related to poverty and urban deprivation, in addition to dependence on heroin. Peter is also clear that the consequences he identifies are primarily social or societal in their origin, including a lack of accommodation, the lack of money or food, and more general discrimination, which are also mostly out of the control of the individual IDU. Such narratives suggest that poverty and urban deprivation play a role in IDUs attitude towards life, death and risk.

Another major theme to arise from the narratives (n = 8) was that death was an occupational hazard of heroin use.

> *Frank, 24 yrs, I think that people who use accept that as one of the risks. You just cop it on the chin.*

> *Joe, 31 yrs, nearly every time, I know its Russian roulette. Sometimes pills. Also some speed, usually hammer first, then speed. Dropping is really an occupational hazard. When your number's up, your number's up. Why worry about it. It's just as likely that you'll have a good whack and then walk across the road and get hit by a truck.*

Finally, not all interviewees exhibited the above-described attitudes towards death and three interviewees reported that they were not indifferent towards death and did their utmost to avoid death.

> *Bruce, 23 yrs, I mean, you talk about friends that have died and that, but I don't really have any sympathy for them. It sounds a bit harsh, but like I say, I've had a lot of friends that have died from one way or the other, you know, but if it's through the choices they made then that's their own business, you know what I mean. I don't want my daughter to know her whole life that her dad died a junkie.*

> *David, 35 yrs, it is out of control in one sense but I don't break into houses or anything like that. The only control I have is to throw myself into an area where it's impossible to get heroin. The best I can do is one day without. There use an element of control I suppose, but it's not enough to break free. I don't want to die. Either that or fail heroicly.*

Worst Consequences

Interviewees were also asked what they believed would be the worst consequence of experiencing an overdose. The interviewees were then read a list of four possible alternatives and asked to nominate one (death, brain damage, police involvement or being woken up). The order of consequences was randomly altered. Interviewees were also able to identify other consequences from which three more responses were identified (nothing, all and wasted money).

Whilst thirteen interviewees reported that death was the worst consequence of overdose, the majority (58%, n = 35) of interviewees identified brain damage as the worst possible consequence of an overdose. Other responses included 'Being woken up' (n = 5, 8.3%) and 'Police Involvement' (n = 3, 5.0%). Whilst this finding is similar to responses from non-IDU populations [39,40], it does demonstrate that the majority of these interviewees clearly identified that there was something worse than death. For example:

Lisa, 25 yrs, I knew a guy who overdosed and ended up with brain damage and he ended up brain dead and they turned the machines off. That was pretty sad really. With my partner, I think about it: is it bad for him to be here brain-dead or with brain damage. I think I'd prefer them to die than have brain damage, but then it's the people that they leave behind. I think a lot of families go through a lot of shit. I mean, it's hard to say. Here I am saying "these families go through a lot of shit," but then I'll go and risk killing myself. For me in an overdose, I'd prefer to die, than have ... brain damage.

The next most common response was 'being woken up.'

Damian, 29 yrs, Coming back with a ... Narcan headache. That's worse than anything I've ever had. I'd definitely rather be dead than brain damaged. It's part of the game isn't it, guaranteed, you're born to die.

Debbie, 22 yrs, for me it was just waking up, that was the pits. For the person overdosing the worst consequence is waking up straight. If you've got people with you, you shouldn't get brain damage. All they're concerned about is the drugs and getting drugs and being stoned.

Discussion

The data presented above highlights that many interviewees did not see the possibility of dying as a reason to reduce risk behaviours. Most experienced the consequences of their risk behaviour regularly, yet few reported engaging in safe using practices. Despite the fact that death is a common occurrence in this group of people and they engage in a behaviour that carries a risk of death every day, most tend to repress their fear of death, treating the likelihood of their death with either ambivalence or indifference. It was apparent that when the effect desired from drug use is on the verge of overdose/death, safe using messages are unlikely to be of sufficient priority. For example, telling an IDU to 'taste' their heroin prior to using the whole amount makes little sense to someone attempting to gain the maximum effect from the heroin they possess. These findings raise questions about the conclusions arising from the existing literature which focuses on changing individual behaviour and suggests support for interventions based on reducing environmental risk [41].

IDUs Relationship with Death

The major finding of the study is the high level of indifference and fatalism displayed by many of the interviewees towa rds their own death and the way in which social and environmental factors such as poverty and marginalisation form the background for this indifference. The narratives support the observations of previous research that many of the interviewees were driven by ambivalent and confused motives [22]. In particular, it was observed that for some IDUs, their death is an event which is viewed with some dispassion and taking measures to try to avoid the death can appear to be the equivalent of 'avoiding the unavoidable.'

The narratives presented also lend weight to the proposal that indifference towards death may turn out to be a rationally based response to "social isolation, meaninglessness and anomie, so characteristic of social life in the 20th century" [[42]: 715]. They point to the reality that in the lived experience of these IDUs where "health may be accorded a relatively low priority by individuals suffering psychological difficulties or social deprivation" [[43]: 223]. Indeed, it is implicit in these narratives that many of the interviewees inhabited a social sphere where there was very little in their lives that supplied meaning apart from substance use, similar to that proposed by Plumridge and Chetwynd [23].

Indifference might also be seen as a matter-of-fact response to the very high death rate amongst heroin users, but can also be viewed as fatalistic. Accepting risk as an 'occupational hazard' may tacitly be denying any sense of agency towards risk behaviour and may result in IDUs not engaging in risk avoidance behaviours. However, the idea of an occupational hazard is common amongst other groups within Western society that engage in high levels of risk behaviour. As seen in Albert's investigation of risk and injury in serious recreational cyclists [18], the concept of occupational hazard is employed widely to deal with situations which, on-the-whole, have little to do with the individual's behaviour and are more related to societal norms surrounding automobile use. The concept of death as an occupational hazard attitude is also reflected in some discourses of soldiers in wartime settings [19,20,44]. However, the literature on attitudes towards death in wartime soldiers emphasises more strongly the conflict-laden nature of such attitudes, particularly in relation to dual roles of masculine provider and citizen. The narratives from interviewees in this study also referred to conflicting elements of their life, most particularly the need to feed their habit while staying alive. From this sample, it was difficult to draw any inferences about gender roles in this regard, although a confrontational attitude towards death was more apparent in men. However, the clearest parallel was the maintenance of the self image in a hostile environment, where heroin use was viewed as a personal behaviour that was made life threatening because of the drug's legal status.

In the context of drug prohibition inhabited by these interviewees, heroin use has an 'unavoidably risk-laden nature' which leaves the IDU no other option than to reasonably accept death as an occupational hazard of heroin use [15,41]. Thus, it appears somewhat incongruous to suggest that IDUs should use in a safe environment when no such environments exist and illustrates the logic of environmental interventions such as safe injecting facilities. On the other hand, it is also evident that some IDU are neither socially marginalised nor will they choose to use such safe environments, and that the intersection between risk, pleasure, escape and indifference means that reducing harm is not a priority.

The narratives of interviewees have also suggested how such societal factors can impact on a person's indifferent state and have illustrated the link between the mental state of the individual, the high cost of heroin and, by association, current drug policy. Similarly, the recognition by the individual that they are realistically unable, to change situations for themselves, ultimately leads to indifferent and fatalistic attitudes towards their own well-being. When individuals are dislodged from the social fabric of society, or have their aspirations consistently thwarted, they are more likely to hold indifferent attitudes towards their death [45]. This was evident in those interviewees who reported being ambivalent towards life as much as being indifferent towards death. Such attitudes have often been documented in socially and economically deprived urban areas [24,25]. It was also reflected in some responses which relayed a sense of bravado towards death and risk, although the process through which individuals developed such responses and the implications that such attitudes have for understanding their attitudes towards death are most probably derived from a combination of an individual interpreting past events in a way which allows maintenance of self image, as well as the telling of their story which reflects the same goal. Far beyond simple epidemiological correlations, ethnographic work has demonstrated how complex economic, social and cultural factors interact to create situations where drugs become a central part [15]. In such a social environment, "people who are using don't care." These findings suggest that beyond investigating and treating drug use, "poverty and deprivation warrant intervention in their own right" [26].

Conclusion

Most of the interviewees in this study were indifferent towards heroin-related death. Whilst interviewees were well aware of the possible consequences of their actions, these consequences were not as important as achieving their desired state of mind. Despite the fact that death is a common occurrence in this group of people and they engage in hazardous behaviour on a daily basis that carries a risk of death, most treat the likelihood of their death with either indifference or resignation. When marginalised groups such as IDUs experience the 'existential angst' observed in some of the narratives presented above, messages of harm reduction and health promotion may be of little relevance.

These findings illustrate that it may be more important to address the reasons behind this indifference than to attempt to change behaviour. In the current drug policy context of prohibition, discourses surrounding the rational choice of IDUs to reduce the risk associated with their drug use are sometimes simplistic and unrealistic. In reality, many IDUs do not have a full range of choices of how to reduce the risks associated with their drug use and the discourses of choice espoused within safe using messages may ultimately fail to serve the drug user and the wider community, encouraging 'victim blaming' thereby further entrenching the marginalisation and fatalism of IDU populations.

Competing Interests

The author declares that they have no competing interests.

Author's Contributions

PGM conducted all elements of this study

Acknowledgements

Special thanks to Associate Professor David Moore for comments on an earlier draft of this paper. I would like to thank Kate Wisbey and Associate Professor Liz Eckermann for their editorial assistance and overall guidance and support, and Richard Marks, formerly of Barwon Health Drug Treatment Services, for all his operational support in conducting this study. This study was funded by a Deakin University Postgraduate Award Scholarship.

References

1. Darke S, Zador D: Fatal Heroin 'overdose': A Review. Addiction 1996, 91(12):1765–72.

2. Harris EC, Barraclough B: Suicide as an outcome for mental disorders. British Journal of Psychiatry 1997, 170:205–228.

3. Dwyer R, Rumbold G: The Illicit Drug Reporting System Project: Community Report. Melbourne: National Drug and Alcohol Research Centre and Turning Point Drug and Alcohol Centre Inc; 1999.

4. Darke S, Ross J, Hall W: Overdose among heroin users in Sydney, Australia: I. prevalence and correlates of non-fatal overdose. Addiction 1996, 91:405–411.

5. Bennett GA, Higgins DS: Accidental overdose among injecting drug users in Dorset, UK. Addiction 1999, 94:1179–1180.

6. Crofts N, Aitken CK, Kaldor JM: The force of numbers: why hepatitis C is spreading among Australian injecting drug users while HIV is not. Medical Journal of Australia 1999, 171:165–166.

7. MacPhail C, Campbell C: 'I think condoms are good but, aai, I hate those things': condom use among adolescents and young people in a Southern African township. Social Science & Medicine 2001, 52:1613–1627.

8. Latkin CA, Forman V, Knowlton A, Sherman S: Norms, social networks, and HIV-related risk behaviors among urban disadvantaged drug users. Social Science & Medicine 2003, 56:465–476.

9. Petersen AR: Risk and the Regulated Self: The Discourse of Health Promotion as Politics of Uncertainty. Australian and New Zealand Journal of Sociology 1996, 32:44–57.

10. Gough B: 'Real men don't diet': An analysis of contemporary newspaper representations of men, food and health. Social Science & Medicine 2007, 64:326–337.

11. Lawlor DA, Frankel S, Shaw M, Ebrahim S, Smith GD: Smoking and ill health: Does lay epidemiology explain the failure of smoking cessation programs among deprived populations? American Journal of Public Health 2003, 93:266–270.

12. Vartiainen E, Korhonen HJ, Koskela K, Puska P: Twenty Year Smoking Trends in a Community-Based Cardiovascular Diseases Prevention Programme: Results from the North Karelia Project. European Journal of Public Health 1998, 8:154–159.

13. Seal KH, Kral AH, Gee L, Moore LD, Bluthenthal RN, Lorvick J, Edlin BR: Predictors and Prevention of Nonfatal Overdose Among Street-Recruited Injection Heroin Users in the San Francisco Bay Area, 1998–1999. Am J Public Health 2001, 91:1842–1846.

14. Maher L, Dixon D, Hall W, Lynskey M: Running the Risks: Heroin, Health and Harm in South West Sydney. Sydney, N.S.W.: National Drug and Alcohol Research Centre, University of New South Wales; 1998.

15. Moore D: Governing street-based injecting drug users: A critique of heroin overdose prevention in Australia. Soc Sci Med 2004, 59:1547–1557.

16. Zador D, Sunjic S, McLennan J: Circumstances and Users' Perceptions of Heroin Overdose at the Time of the Event and at One-Week Follow-Up in Sydney, Australia: Implications for Prevention. Addiction Research & Theory 2001, 9:407–423.

17. Alexander M, Lester D: Fear of death in parachute jumpers. Perceptual and Motor Skills 1972, 34:338.

18. Albert E: Dealing with Danger: The Normalization of Risk in Cycling. International Review for the Sociology of Sport 1999, 34:157–171.

19. Sasson-Levy O: Military, masculinity, and citizenship: Tensions and contradictions in the experience of blue-collar soldiers. Identities 2003, 10:319–345.

20. Britton D, Williams C: " Don't Ask, Don't Tell, Don't Pursue": Military Policy and the Construction of Heterosexual Masculinity. J Homosex 1995, 30:1–22.

21. Plumridge E, Chetwynd J: The moral universe of injecting drug users in the era of AIDS: sharing injecting equipment and the protection of moral standing. AIDS Care 1998, 10:723–734.

22. Neale J: Suicidal intent in non-fatal illicit drug overdose. Addiction 2000, 95:85–93.

23. Plumridge E, Chetwynd J: Identity and the social construction of risk. Sociology of Health and Illness 1999, 21:329–343.

24. Bourgois P: Search of Respect: Selling Crack in El Barrio. 2nd edition. Cambridge: Cambridge University Press; 2003.

25. Allen C: The poverty of death: social class, urban deprivation, and the criminological consequences of sequestration of death. Mortality 2007, 12:79–93.

26. Valentine K, Fraser S: Trauma, damage and pleasure: Rethinking problematic drug use. International Journal of Drug Policy 2008, 19:410–416.

27. Aldridge J, Parker H, Measham F: Drug Trying and Drug Use Across Adolescence: A Longitudinal Study of Young People's Drug Taking in Two Regions of Northern England. London: Home Office/Drugs Prevention Advisory Service; 1999.

28. Parker H, Bury C, Egginton R: New Heroin Outbreaks Among Young People in England and Wales. In Crime Detection and Prevention Series Paper 92. London: Home Office; 1998.

29. Miller PG: Dancing with Death: The Grey Area between Suicide Related Behaviour, Indifference and Risk Behaviours of Heroin Users. Contemporary Drug Problems 2006, 33:427–453.

30. Miller P: Dancing with Death: Risk, Health Promotion and Injecting Drug Users. PhD thesis. Deakin University, School of Social Inquiry; 2002.

31. Miller PG: Scapegoating, Self-confidence and Risk Comparison: The Functionality of Risk Neutralisation and Lay Epidemiology by Injecting Drug Users. International Journal of Drug Policy 2005, 16:246–253.

32. Kellehear A: The Unobtrusive Researcher: A Guide to Methods. St. Leonards, NSW, Australia: Allen & Unwin; 1993.

33. Pope C, Mays N: Qualitative Research: Reaching the parts other methods cannot reach: an introduction to qualitative methods in health and health services research. BMJ 1995, 311(1 July):42–45.

34. Stenius K, Mäkelä K, Miovsky M, Gabrhelik R: How to Write Publishable Qualitative Research. [http://www.parint.org/isajewebsite/isajebook2.htm] In Publishing Addiction Science: A Guide for the Perplexed Second edition. Edited by: Babor TF, Stenius K, Savva S. Rockville, MD: International Society of Addiction Journal Editors; 2008, 82–97.

35. Des Jarlais DC, Lyles C, Crepaz N, TREND Group: Improving the Reporting Quality of Nonrandomized Evaluations of Behavioral and Public Health Interventions: The TREND Statement. Am J Public Health 2004, 94:361–366.

36. Darke S: Self-report among injecting drug users: A review. Drug and Alcohol Dependence 1998, 51:253–263.

37. Rhodes T, Cusick L: Love and intimacy in relationship risk management: HIV positive people and their sexual partners. Sociology of Health & Illness 2000, 22:1–26.

38. Martin A, Stenner P: Talking about drug use: what are we (and our participants) doing in qualitative research? International Journal of Drug Policy 2004, 15:395–405.

39. Florian V, Mikulincer M: Fear of death and the judgment of social transgressions: A multidimensional test of terror management. Journal of Personality & Social Psychology 1997, 73:369–381.

40. Lester D: Fear of death in suicidal persons. Psychological Reports 1967, 20:1077–1078.

41. Rhodes T: The 'risk environment': a framework for understanding and reducing drug-related harm. International Journal of Drug Policy 2002, 13:85–94.

42. Kellehear A: Are we a 'death-denying' society? a sociological review. Social Science and Medicine 1984, 18:713–723.

43. Kelly M, Charlton B: The Modern and the Postmodern in Health Promotion. In The Sociology of Health Promotion: Critical Analyses of Consumption,

Lifestyle and Risk. Edited by: Bunton R, Nettleton S, Burrows R. New York, NY: Routledge; 1995:79–91.

44. Small N: Death and difference. Death, gender and ethnicity 1997, 202.

45. Travis R: Suicide in Cross-Cultural Perspective. International Journal of Comparative Sociology 1990, 31:3–4.

Factors Associated with Attitudes Towards Intimate Partner Violence Against Women: A Comparative Analysis of 17 Sub-Saharan Countries

Olalekan A. Uthman, Stephen Lawoko and Tahereh Moradi

ABSTRACT

Background

Violence against women, especially by intimate partners, is a serious public health problem that is associated with physical, reproductive and mental health consequences. Even though most societies proscribe violence against women, the reality is that violations against women's rights are often sanctioned under the garb of cultural practices and norms, or through misinterpretation of religious tenets.

Methods

We utilised data from 17 Demographic and Health Surveys (DHS) conducted between 2003 and 2007 in sub-Saharan Africa to assess the net effects of socio-demographic factors on men's and women's attitudes toward intimate partner violence against women (IPVAW) using multiple logistic regression models estimated by likelihood ratio test.

Results

IPVAW was widely accepted under certain circumstances by men and women in all the countries studied. Women were more likely to justify IPVAW than men. "Neglecting the children" was the most common reason agreed to by both women and men for justifying IPVAW followed by "going out without informing husband" and "arguing back with the husband." Increasing wealth status, education attainment, urbanization, access to media, and joint decision making were associated with decreased odds of justifying IPVAW in most countries.

Conclusion

In most Sub-Saharan African countries studied where IPVAW is widely accepted as a response to women's transgressing gender norms, men find less justification for the practice than do women. The present study suggests that proactive efforts are needed to change these norms, such as promotion of higher education and socio-demographic development. The magnitude and direction of factors associated with attitudes towards IPVAW varies widely across the countries, thus suggesting the significance of capitalizing on need-adapted interventions tailored to fit conditions in each country.

Background

Intimate partner violence against women (IPVAW) is deep-rooted in many African societies, where it is considered a prerogative of men [1,2] and a purely domestic matter in the society [3,4]. IPVAW is one of the greatest barriers to ending the subordination of women. Women, for fear of violence, are unable to refuse sex or negotiate safer sexual practices, thus increasing their vulnerability to HIV if their husband is unfaithful [5,6]. Violence against women, especially by intimate partners, is a serious public health problem that is associated with physical, reproductive and mental health consequences [7-10]. Even though most societies proscribe violence against women, the reality is that violations against women's rights are often sanctioned under the garb of cultural practices and norms, or through misinterpretation of religious tenets. Moreover, when violation takes place within the home, as it is often the case, the abuse is effectively ignored by the tacit silence and the passivity displayed by the state and the law-enforcing machinery. The global dimensions of this violence are alarming as highlighted by numerous studies [2,7,8,11-25].

A troubling aspect of IPVAW is its benign social and cultural acceptance of physical chastisement of women and isthe husband's right to "correct" an erring wife [26]. Women's susceptibility to IPVAW has been shown to be greatest in societies where the use of violence in many situations is a socially accepted norm[27]. Studies have shown that attitude towards IPVAW is one of the most prominent predictors of IPVAW, when contrasted with other potential predictors including social and empowerment factors [28-30]. As it has been emphasised by a number of scholars [31-33], without a fundamental change in the social attitudes that foster, condone, and perpetuate IPVAW we will not be able to respond effectively to this problem, by substantially reducing its alarming rates. Women's own condemnation of this behaviour may, therefore, be an important element in changing it. Most of the studies in the low- and middle-income countries on IPVAW have focused on actual prevalence of IPVAW and its determinants [2,7,8,11-25] and less focus has been on the underlying attitudes towards IPVAW [5,34-38]. In addition, most of the existing studies on IPVAW are based on women's responses while men's perspective may also play an important role. Knowing the extent and reasons for justification of IPVAW in a particular setting is important for different reasons[35]. First, unfettered social and cultural acceptance of IPVAW may not only lead to abetting such practices, but may also create major obstacles toward altering such practices. Hence, understanding the underlying factors related to positive attitude towards IPVAW may be fundamental for designing effective programmes to address the issue. Second, acceptance of IPVAW can be considered as an indicator of the status of women in a specific social and cultural setting. Levels of acceptance of IPVAW can provide insights into the stage of social, cultural and behavioural transformation of a specific society in its evolution towards a more gender egalitarian society.

Conceptual Framework and Hypotheses

Building largely upon conceptual framework developed by Rani and colleagues [35], we postulated that an important 'trigger' for IPVAW in patriarchal societies is the transgression of established gender roles. Studies from different patriarchal societies have identified a common set of role expectations for women including preparing food properly, caring for children, seeking husband's or other family member's permission before going out, not arguing with husband, and meeting the sexual needs of the husband [5,35-38]. The conceptual framework used in this study is based on the social learning theory and the ecological framework in order to understand the predictors of attitude towards IPVAW. Social learning theory postulates that individuals learn how to behave by observing and re-enacting the behaviour of role models. Social norms and gender roles in a patriarchal society are learned within a social group and transmitted from generation to generation. The myth of male superiority is maintained in many societies through rigid gender norms and social practices such as polygamy, restriction on movement of women, bride price and other practices that result in overall lower achievement levels among women including education, employment, financial power, public role. Factors that will promote intolerant attitudes towards IPVAW will operate mainly via three mechanisms [35]: by producing a conflict between reality and myth of male superiority; by exposing people to more egalitarian social networks and authority structures other than kin-based ones; and by exposing to non-conformist ideas through modern media. Wealth defines class, which may be characterised by different social networks. Since poverty may increase chances of conflict over resources, it is likely that individuals growing up in poor households and neighbourhoods are often exposed to violence both within and outside the family resulting in high acceptance of violence to resolve conflicts. Furthermore, education and urbanisation may have a greater inverse effect on acceptance of IPVAW among women than among men. The purpose of this study was to contribute to the growing empirical literature on attitudes towards IPVAW. The specific objectives were two-folded; 1) to study gender differences in men's and women's attitudes towards IPVAW and 2) to examine factors associated with attitudes towards IPVAW.

Methods

Data

This study used data from Demographic and Health Surveys (DHS) conducted between 2003 and 2007 in sub-Saharan Africa available as of November 2008. DHS surveys were implemented by respective national institutions and ORC Macro International Inc. with financial support from the US Agency for International Development. Methods and data collection procedures have been published elsewhere [39]. Briefly, DHS data are nationally representative, cross-sectional, household sample surveys with large sample sizes, typically between 5,000 and 15,000 households. The sampling design typically involves selecting and interviewing separately nationally representative probability samples of women aged

15–49 years and men aged 15–59 years based on multi-stage cluster sampling, using strata for rural and urban areas and for different regions of the countries. A standardized questionnaire was administered by interviewers to participants in each country. The survey's questionnaireswere similar across countries yielding inter-country comparable data. Only countries with available data on attitudes towards IPVAW were included in this study. This resulted in inclusion of the following 17 participating countries in DHS: Benin, Burkina Faso, Ethiopia, Ghana, Kenya, Lesotho, Liberia, Madagascar, Malawi, Mozambique, Namibia, Nigeria, Rwanda, Swaziland, Tanzania, Uganda and Zimbabwe.

Outcome Variable

To assess the degree of acceptance of IPVAW by women and men, respondents were asked the following question: "Sometimes a husband is annoyed or angered by things which his wife does. In your opinion, is a husband justified in hitting or beating his wife in the following situations?" The five scenarios presented to the respondents for their opinions were: 1. "if wife burns the food," 2. "if wife argues with the husband," 3. "if wife goes out without informing the husband," 4. "if wife neglects the children," and 5. "if the wife refuses to have sexual relations with the husband." Information was collected from all women and men irrespective of their marital status. A binary outcome variable was created for acceptance of IPVAW, coded as '0' if the respondent did not agree with any of the situations when a husband is justified in beating the wife or did not have any opinion on the issue and coded as '1' if the respondent agreed with at least one situation where the husband is justified in beating the wife.

Determinants Variables

To assure consistency, we selected determinant variables based upon previous studies that investigated factors associated with attitudes towards IPVAW.

Demographic/social position was assessed using the following indicators: Sex of respondent was defined as men or women; age (15–24, 25–24, 35+ years), place of residence (urban or rural area), occupation (working or not working), education (no education, primary, secondary or higher), marital status (never, currently, or formerly married). DHS did not collect direct information on household income and expenditure. We used DHS wealth index as a proxy indicator for socioeconomic position. The methods used in calculating DHS wealth index have been described elsewhere [40-42]. Briefly, an index of economic status for each household was constructed using principal components analysis based on the following households' variables: number of rooms per house, ownership of car, motorcycle, bicycle, fridge, television and telephone, and kind of heating device. From this the DHS wealth index quintiles (poorest, poor, middle, rich, and richest) were calculated and used in the subsequent modelling.

Media access was assessed using the following indicators: access to information measured via frequency of watching television, listening to radio, and reading newspapers/magazine. To allow meaningful statistical analysis, we dichotomized the response levels "less than one week," "at least once a week," and "almost every day" as one group and the response level "not at all" as the other group.

Decision Making Power

Respondents' decision autonomy were assessed by inquiring about who bore the responsibility of making decisions on household purchases including small and large ones, visiting relatives and friends, spending the wife's earnings, and the number of children to have. For these variables, response options were "husband," "wife," or "both husband and wife." We created set of additive scale (from 0 to 5) that counted the number of domains in which each (husband/partner alone, wife alone, and couple) had the final word.

Statistical Analyses

In the descriptive statistics the distribution of respondents by the key variables were expressed as percentages. We used Pearson's chi-squared test for analyzing contingency tables. All cases in the DHS data were given weights to adjust for differences in probability of selection and to adjust for non-response. Individual weights were used for descriptive statistics in this study. We used multiple logistic regressions to examine factors associated with attitudes towards IPVAW. We entered all covariates simultaneously in the multiple regression models. Results were presented in the form of odds ratio (ORs) with significance levels and 99% confidence intervals (99% CIs). We performed random-effects estimates models as described by DerSimonian and Laird [43] to incorporate between-country heterogeneity in addition to sampling variation for the calculation of summary OR estimates and corresponding 99% CIs. Between countries heterogeneity was assessed using the Cochran Q test [44] and the I^2 statistic [45], which describes the percentage of total variation across countries that is the result of heterogeneity rather than chance. I^2 was calculated based on the formula $I^2 = 100\% \times (Q - \text{degree of freedom})/Q$.

Regression diagnostics were used to judge the goodness-of-fit of the model. They included the tolerance test for multicollinearity, its reciprocal variance inflation factors (VIF), presence of outliers and estimates of adjusted R square of the regression model.

The largest VIF greater than 10 or the mean VIF greater than 6 represent acceptable fit of the models [46,47]. Statistical methods for complex survey data, Stata, release 10.0 (Stata Corp., College Station, TX, USA) were used to account for stratification, clustered sampling and weighing to estimate efficient regression coefficients and robust standard errors. All tests were two tailed. Since due to the large sample size, small differences in attitudes between groups may easily reach the conventional 0.05 statistical significance, we reduced the condition for significance to 0.01 to account for this effect.

Ethical Consideration

This study is based on an analysis of existing survey data with all identifier information removed. The survey was approved by the Ethics Committee of the ORC Macro at Calverton in the USA and by the National Ethics Committee in the respective country. All study participants gave informed consent before participation and all information was collected confidentially.

Results

Description of Included Countries

Table 1 shows the countries, years of data collection, and sample sizes. It also illustrates the demographic and economic diversity of the selected countries. All the 17 countries were low-income countries. As for gross domestic product (GDP) per capita, Swaziland and Namibia emerged as the most affluent countries with values higher than United States dollar (US$)2000 per capita, whilst by contrast Ethiopia, Malawi and Rwanda were the most deprived with values less than US$250 per capita. Nigeria was the most and Lesotho was the least populated country among the countries studied. Regarding levels of urbanization, the percentage of urban population varied across the countries. The percentage of literacy among women was highest in Lesotho (90%) and lowest in Burkina Faso (17%).

Table 2 shows the socio-demographic characteristics of the study participants. Most of the respondents were female. The percentage of female ranged from 53% in Ghana to 81% in Mozambique. Most of the respondents (34% to 48%) were aged 15–24. The percentage of respondents with no education varies across the country. The percentage of respondents with no education was lowest in Zimbabwe (3%) and highest in Burkina Faso (76%). With the exception of Lesotho (44%), more than 50% of the respondents were currently working. The percentage of respondents that are currently married ranged from 35% in Namibia to 73% in Benin. Respondents were fairly evenly distributed across the wealth status strata. In most countries, most of the respondents were living in the rural areas.

Burkina Faso (11%) had least number of respondents with access to newspaper; Swaziland (68%) had the highest. In all countries studied, more than 50% had access to radio. The percentage of respondents with access to television ranged from 17% in Malawi to 59% in Ghana.

Table 1. Description of data sets, selected social, economic, and demographic characteristics of the countries included in the study

Variable	year	Sample size		Population			GDP per capita		Adult literacy rate	
		Men	Women	Total (millions 2005)	Growth rate (1975 – 2005)	%urban (2005)	Value (US$ 2005)	Growth rate (1990 – 2005)	Men	Women
Benin	2006	6000	18000	8.5	3.2	40.1	508	1.4	47.9	23.3
Burkina Faso	2003	3605	12477	13.9	2.8	18.3	391	1.3	31.4	16.6
Ethiopia	2005	6033	14070	79	2.8	16	157	1.5	50.0	22.8
Ghana	2003	5015	5691	22.5	2.6	47.8	485	2.0	66.4	49.8
Kenya	2003	3578	8195	35.6	3.2	20.7	547	-.1	77.7	70.2
Lesotho	2004	2797	7095	2.0	1.8	18.7	808	2.3	73.7	90.3
Liberia	2007	6009	7092	3.4	2.5	58.1	167	2.3	58.3	45.7
Madagascar	2004*	2432	7949	18.6	2.9	26.8	271	-.7	76.5	65.3
Malawi	2004	3261	11698	13.2	3.1	17.2	161	1.0	74.9	54.0
Mozambique	2003	2900	12418	20.5	2.2	34.5	335	4.3	54.8	25.0
Namibia	2007	3915	9804	2.0	2.7	35.1	3016	1.4	86.8	83.5
Nigeria	2003	2346	7620	141.4	2.8	48.2	752	0.8	78.2	60.1
Rwanda	2005	4820	11321	9.2	2.5	19.3	238	0.1	71.4	59.8
Swaziland	2006	4156	4987	1.1	2.5	24.1	2414	0.2	80.9	78.3
Tanzania	2004	2635	10329	38.5	2.9	24.2	316	1.7	77.5	62.2
Uganda	2006	2503	8531	28.9	3.3	12.6	303	3.2	76.8	57.7
Zimbabwe	2006*	7175	8907	13.1	2.5	35.9	259	-2.1	92.7	86.2

(Source: †Demographic and Health Surveys of the respective countries; ‡UNDP Human Development Report)
GDP: gross domestic product, HDI: human development index, GDI: gender-related development index

Table 2. Percentage distribution by selected characteristics

Variable	Benin %	Burkina Faso %	Ethiopia %	Ghana %	Kenya %	Lesotho %	Liberia %	Madagascar %	Malawi %	Mozambique %	Namibia %	Nigeria %	Rwanda %	Swaziland %	Tanzania %	Uganda %	Zimbabwe %
Sex																	
Men	23.0	22.4	30.0	46.8	38.4	28.3	45.9	23.4	21.8	18.9	38.5	23.5	29.9	46.5	20.3	22.7	44.6
Women	77.0	77.6	70.0	53.2	69.6	71.7	54.1	76.6	78.2	81.1	71.5	76.5	70.1	54.5	79.7	77.3	55.4
Age																	
15–34	32.9	48.3	40.7	36.4	42.8	46.3	37.9	36.3	43.2	40.8	41.8	41.2	43.3	48.4	41.6	41.6	46.6
25–34	34.0	27.4	29.3	29.4	29.3	25.3	29.3	30.5	31.6	39.5	30.8	29.3	27.1	26.8	31.5	30.3	28.4
35+	32.1	22.3	29.9	34.2	27.9	29.4	32.8	33.2	25.2	29.6	27.4	29.6	29.7	24.8	26.9	28.1	25.1
Education																	
No education	58.8	76.2	54.2	28.3	13.5	67.3	29.8	14.6	28.7	32.0	08.6	35.1	21.2	08.1	22.0	17.4	83.1
Primary	21.9	13.2	24.4	18.4	53.3	58.8	34	39.8	62.6	56.6	27.7	22.8	67.2	33.5	64.0	59.1	31.6
Secondary+	19.2	10.6	21.4	53.3	33.3	23.9	36.2	45.6	16.7	11.3	63.7	42.1	11.6	58.3	14.0	23.5	65.3
Occupation																	
Working	80.5	86.7	50.5	79.2	64.9	44.9	71.1	73.2	63.6	72.9	56.9	61.2	67.9	50.7	79.1	88.1	54.6
Not working	19.5	13.3	49.5	20.8	35.1	55.1	28.9	34.8	36.4	27.1	43.1	38.8	32.1	49.3	20.9	11.9	45.4
Marital																	
Never married	22.6	34.9	31.3	32.8	34.4	38.0	32.6	26.3	19.7	21.1	58.7	31.9	40.5	56.5	28.2	26.9	36.7
Currently married	73.1	71.6	59.6	60.0	57.2	49.9	59.8	61.9	70.2	66.3	35.0	63.7	49.2	36.6	63.0	61.7	52.8
Formerly married	04.3	03.5	09.1	07.2	8.4	12.1	7.5	11.8	10.2	12.6	06.3	04.4	10.3	06.9	08.8	11.4	10.5
Wealth																	
Poorest	18.7	16.7	19.7	23.9	16.3	17.2	19.3	12.5	16.7	18.9	15.7	19.1	19.5	14.9	17.4	20.5	17.8
Poor	19.1	18.5	15.0	18.2	15.8	19.8	19.8	10.7	20.1	15.1	17.4	18.0	18.7	16.4	18.5	19.0	18.5

Table 2. *(Continued)*

Middle	19.3	22.1	14.6	17.5	17	18.3	19.8	13.1	21.8	17.6	34.1	19.6	18.1	18.8	18.1	17.5	18.2
Richer	21.0	16.4	14.3	18.6	19.7	20.5	20.7	18.7	21.3	30.7	25.1	20.8	19.6	21.7	22.7	18.3	23.0
Richest	21.8	25.2	36.5	21.8	31.2	34.2	20.3	44.9	20.1	27.6	17.7	22.5	34.1	28.2	23.3	34.7	22.5
Type of residence																	
Urban	42.0	24.5	30.1	39.9	33.1	26.7	43.7	64.3	14.3	43.7	44.3	40.6	23.2	32.6	24.0	16.7	35.2
Rural	58.0	75.5	69.9	60.1	66.9	73.3	56.3	35.7	85.7	56.3	53.7	59.4	76.8	67.4	76.0	83.3	64.8
Decision making indices																	
Respondent alone (1–5)	58.2	62.6	42.9	59.1	73.6	70.2	53.1	65.2	66.4	73.3	31.2	52.7	62.4	16.2	78.3	45.7	16.0
Husband/Partner alone (1–5)	43.5	57.5	42.9	29.6	52.7	46.0	45.4	23.9	50.5	51.0	30.1	56.8	28.8	16.9	54.0	55.7	10.9
Husband-wife (1–5)	35.1	19.3	58.4	24.6	42.7	38.6	59.9	52.2	42.6	42.0	43.5	29.9	41.1	14.9	36.7	46.8	28.1
Media access																	
Read newspaper	14.1	11.8	26.5	28.8	31.1	23.6	32.9	43.7	32.8	19.0	66.6	30.4	29.4	68.2	42.1	28.2	50.3
Listen to radio	84.5	72.6	53.7	89.6	85.6	58.9	70.5	79.1	83.0	86.8	89.1	78.7	81.4	83.6	80.7	82.7	64.3
Watch television	40.8	30.2	32.3	59.0	46.4	18.0	44.1	47.2	17.1	36.7	51.5	49.8	17.1	48.4	23.2	18.3	44.7

Justification of IPVAW by Gender Norm Transgressed

"Neglecting the children" was the most common reason agreed by both women (Figures 1) and men (Figures 2) for justifying IPVAW followed by going out without informing husband and arguing back with the husband. The proportion of respondents who agreed with the statement that IPVAW is justified for "neglect-

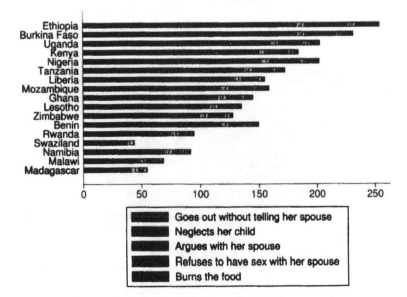

Figure 1. Percentage of women who believe that IPVAW is justified, by different scenarios.

ing the children" ranged from 5% in Madagascar to 49% in Kenya among men and from 11% in Swaziland to 59% in Ethiopia among women. The justification for IPVAW was relatively low for "refusing sexual relations" among scenarios presented. Women were consistently more likely to justify IPVAW than men in all the countries, with the exception of Lesotho, Swaziland and Kenya (Figure 3). The percentage of women who justified IPVAW was lowest in Madagascar (28%) and highest in Ethiopia (74%). Madagascar had also the lowest percentage (8%) of men who justified IPVAW and Kenya the highest (62%).

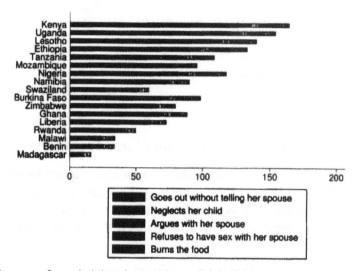

Figure 2. Percentage of men who believe that IPVAW is justified, by different scenarios.

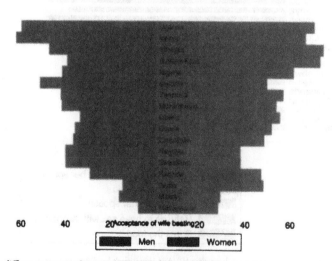

Figure 3. Sex-difference in attitude toward IPVAW of 17 sub-Saharan countries.

Factors Associated with Attitudes Towards IPVAW

Table 3 presents the adjusted OR for justification of IPVAW (see additional file 1 for full odds ratios, 99% CI and p-values). The diagnosis of multi-collinearity is shown in additional file 1. The largest VIF ranged from 2.15 to 5.18; and the average VIF ranged from 1.77 to 2.49. Since none of the VIF values exceeds 10 and none of the average VIF exceeds 6, we concluded that there was no multi-collinearity problem. Women were significantly more likely to justify IPVAW than men in all countries studied with the exception of Lesotho. Women were 29% less likely to justify IPVAW than men in Lesotho (OR = 0.71, 99% CI 0.60 – 0.84). The association between sex and justification of IPVAW became non-significant in Namibia, Kenya, and Swaziland after controlling for respondents' socio-demographic factors, decision making autonomy, and access to media. Compared to respondent aged 35 and older, respondent aged 15–24 were consistently and significantly more likely to justify IPVAW in all countries except for Benin and Burkina Faso. Lower educational attainment was positively associated with acceptance of IPVAW. Respondents with no education or primary education were more likely to justify IPVAW compared with those with secondary or higher education in all countries but Liberia, Madagascar, and Nigeria. Relationship between occupation and acceptance of IPVAW was mixed. Respondents not in working force from Burkina Faso, Mozambique and Rwanda were at 20% statistically increased risk of justifying IPVAW. Currently not working respondents from Benin, Liberia, Madagascar, Malawi, Tanzania, and Zimbabwe were less likely to justify IPVAW. The association was not significant in other seven countries.

Table 3. Factors associated with attitudes towards intimate partner violence against women identified by multiple logistic regression analyses*

Compared with those never married, respondents that were currently married from Benin (OR = 1.44, 99% CI 1.19 – 1.75), Kenya (OR = 1.46, 99% CI 1.21 – 1.77), and Madagascar (OR = 1.35, 99% CI 1.02 – 1.77) were more likely to justify IPVAW. While, those currently married from Malawi, Namibia, Rwanda, and Zimbabwe were less likely to justify IPVAW than those never married. In some countries, such as Benin, Burkina Faso, Ethiopia, Kenya, and Liberia those formerly married were more likely to justify IPVAW. In other countries, such Malawi (OR = 0.56, 99% CI 0.43 – 0.72), Rwanda (OR = 0.78, 99% CI 0.64 – 0.96), and Tanzania (OR = 0.78, 99% CI 0.63 – 0.97) those formerly married were less likely to justify IPVAW than never married. The odds of justifying IPVAW increased with decreasing wealth status in all countries. Living in rural areas increased the odds of justifying IPVAW in most of the countries. However, those living in rural areas in Madagascar (OR = 0.73, 99% CI 0.62 – 0.86) were less likely to justify IPVAW than their counterparts from urban areas. Association of justifying IPVAW with decision making indices were not consistent across the countries studied. Respondents who reported final say in more household decisions than their partners were more likely to justify IPVAW in nine countries and less likely to justify IPVAW in Benin, Burkina Faso, and Mozambique. Respondents were more likely to justify IPVAW in most countries when their partners alone had the final say in more household decisions than they did. When respondents reported more decisions being made jointly than individually, they were significantly less likely to justify IPVAW in most countries.

Access to newspaper reduced the odds of justifying IPVAW in all countries with the exception of Malawi (OR = 1.20, 99% CI 1.06 – 1.37). The association between listening to radio and acceptance of IPVAW was significant in only three countries. As expected, listening to radio reduced the odds of justifying IPVAW in Madagascar (OR = 0.83, 99% CI 0.70 – 0.99) and Rwanda (OR = 0.80, 99% CI 0.72 – 0.90). Counter intuitively, access to radio increased the likelihood of justifying IPVAW in Zimbabwe (OR = 1.23, 99% CI 1.10 – 1.37). The association between watching television and odds of justifying IPVAW was not consistent across countries. In some countries, such as Ethiopia, Ghana, and Madagascar watching television reduced the likelihood of justifying IPVAW. In other countries, such as Mozambique, Namibia and Tanzania watching television increased the odds of justifying IPVAW.

Figure 4 shows the results of pooled odds ratios (weight average) of the determinants of attitudes towards IPVAW (see additional file 2 for forest plots for each variable). The results of meta-analyses confirmed that sex, age, education attainment, wealth status, when partner alone had the final say in household decisions, and access to newspaper were associated with attitudes towards IPVAW in the pooled analyses. Random effect model meta-analysis showed that women were more likely to justify IPVAW than men (pooled OR = 1.98, 99% CI 1.32 to 2.80). The results from the pooled analyses also confirmed that odds of justifying IPVAW increase with decreasing age, decreasing education attainment, decreasing wealth status. Compared those living in the urban areas, those from rural were more likely to justify IPVAW (pooled OR = 1.15, 99% CI 1.02 to 1.30). Random effect model meta-analysis showed respondent were more likely to justify IPVAW when their partners alone had the final say in more household decision that they did (pooled weighted average OR = 1.13, 99% CI 1.08 to 1.18). The pooled OR for the effect of access to newspaper was 0.85 (99% 0.79 to 0.92). The results of pooled analyses for occupation, marital status, when respondents reported more final say in more, when respondents reported more decisions being made jointly, access to radio and television were not significant. Figure 4 also shows magnitude of cross countries variability in the determinants of attitudes towards IPVAW. The Cochran Q's test for heterogeneity for all variables gave p-values which were highly significant (p <.0001). Higgins and Thompson statistics suggested that 79% to 99% of the total variation in the estimated effect of determinants was due to heterogeneity between countries, thus suggesting that between countries heterogeneity were almost certain present.

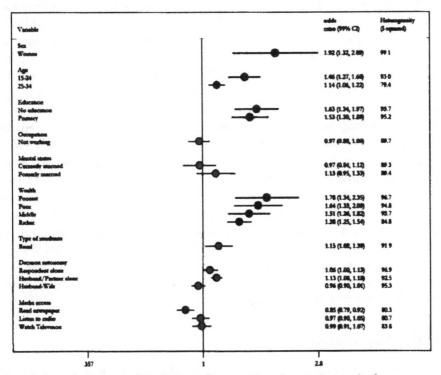

Figure 4. Forest plot showing pooled odds ratio and 99% confidence for socio-demographic factors.

Discussion

In this large comparative study from 17 countries in sub-Saharan Africa, we found that IPVAW was widely acceptable under certain circumstances and more such among women, younger people, less educated, poorest, those living in rural areas, those with less access to media and single decision makers. Women were more likely to justify IPVAW in all countries, with exception of Lesotho even after controlling for confounding factors. This, in agreement with the result of previous study that has examined this association in seven countries in sub-Saharan Africa conducted between 1999 and 2001[35]. A possible explanation for the exception seen in Lesotho could be due to the fact that the adult female literacy rate is higher than adult male literacy rate. It has been reported that women are more vulnerable to abuse and exploitation in environment where there is high gender inequality, these factors may be responsible for the observed gender disparities in attitudes toward IPVAW. More sophisticated measures such as decomposition analyses are needed to explore the sources of gender disparities in attitudes towards IPVAW.

Evidence from meta-analyses suggests that sex of the respondent stood out as the most important predictor of attitudes towards IPVAW. Results of meta-analyses provided evidence that wealth status and education attainment were also significantly associated with attitudes IPVAW. Access to newspaper reduced likelihood of having tolerant attitudes towards IPVAW. Some of the socio-demographic correlates we studied have been documented in literature [12-20]. We found that younger people are more likely to justify IPVAW. Despite the cross-sectional design in this study, comparing trend in attitude by ages indicates that the younger generation is more likely to accept IPVAW than the older one. However, there is a need for longitudinal studies to confirm this finding. Wealth status, education and urbanisation had a greater negative impact on acceptance of IPVAW in most countries in this study. The limited effects seen in primary education alone compared to those with secondary or higher education is not surprising. Having few years of education usually at young age may not expose people to new non-conformists ideas[35]. It may even bring conflict between reality and myth of male superiority[35]. In accordance with the results of previous studies[5,35], we found that occupation status had minimal effect on acceptance of IPVAW. Most women in low-income countries work largely in informal sectors with low paid jobs. Women are usually exposed to the same patriarchal social structures at the work place that may further strengthen the myth of male superiority.

Policy Implications

We have provided evidence that in most Sub-Saharan African countries studied here IPVAW is widely accepted as a response to women's transgressing gender norms, men find less justification for the practice than do women. We believe like others [35,48], that the first step toward eliminating this practice is to "build up a substantial amount of momentum" in opposition to the use of violence in conflict resolution and that, given its widespread acceptance in these societies, the development of a "new social consensus" albeit a slow process, is crucial. A climate of tolerance of IPVAW would make it easier for perpetrators to persist in their violent behaviour and make it more difficult for women to disclose domestic violence [31]. In terms of IPVAW, there is a need for a social environment characterized by low tolerance and an increased sense of social and personal responsibility toward IPVAW [33]. This, in turn, would contribute to a social environment more effective in terms of social control of IPVAW [49]. Public awareness and education campaigns aiming to lowering social tolerance and to increase the sense of social and personal responsibility toward IPVAW are needed in order to reduce and prevent IPVAW.

Sub-Saharan Africa is ethnically, culturally and religiously diverse and economic development and education levels vary widely across the countries. Not unexpectedly, we found that the magnitude and directions of factors associated with attitudes towards IPVAW varies widely across the 17 countries studied. Sub-Saharan African countries have heterogeneous conditions. Understanding cross countries diversities may aid in the identification of regions that may need to be particularly targeted with education and prevention programs. Thus, multifaceted geographically differentiated intervention may represent a potentially effective approach for addressing issues related to intimate partner violence in sub-Saharan Africa with policies tailored to country-specific conditions. Furthermore, decision makers should capitalize on need-adapted interventions to meet societal conditions in a bid to change men's distorted attitudes toward IPVAW.

Potential public health programmes could include structural and gender-based interventions. Structural interventions focusing on improving the coverage and dissemination of information to the general public may be beneficial in changing men's attitudes toward IPVAW, alongside a review of the educational system, which may seem to reinforce gender inequity. It is also important to note that access to media reduced odds of acceptance of IPVAW in most countries. The widespread acceptance of IPVAW may also become a major hurdle in success of other reproductive health programs (i.e., family planning programs), care seeking for sexually transmitted diseases or voluntary testing and counselling, and condom use for prevention of HIV/AIDS if the women do not confront men because of the threat of domestic violence, as a large proportion of women in these societies considered "arguing with husband" and "refusing sex" as valid reasons for wife beating[50]. Gender-based interventions, building on advocacy for shared autonomy in the domestic domain, and the provision of basal education for all may prove paramount in changing men's distorted attitudes about IPVAW, particularly among younger men and in rural settings. We found that joint decision making reduced likelihood of justifying IPVAW indicating that imbalance of power is associated with higher odds of justifying IPVAW. Interventions that promote joint decision-making might be a promising strategy for increasing women's view towards equality in marriage while promoting men's views that household disputes should be settled with negotiation and not violence. To break the norms that sustain women's vulnerability in society, there is a need for pro-active efforts toward socioeconomic development and promotion of higher education.

Study Limitations and Strengths

There are a number of caveats to be considered when interpreting these results. The cross-sectional nature of the data limits ability to draw casual inferences. The study can be criticized for using an indirect measure of household wealth. However, due to the fact that in low- and middle-income countries, it is hard to obtain reliable income and expenditure data, an asset-based index is generally considered a good proxy for household wealth status. Our study focused on understanding the role of individual variables as determinants of attitudes towards violence in specific Sub-Saharan countries. We did not incorporate an assessment of the effect of interactions between such variables and other societal factors in our study design. For instance societal level variables such as ethnicity could interact with gender in explaining attitudes towards violence. Future research using a multilevel design may be necessary to assess such interaction effects. Another important limitation is that the reliability and validity of this instrument used for measuring attitudes towards IPVAW is yet to be established [36,37]. It has been documented that attitudes toward IPVAW is limited in scope to capture women's normative roles in the domestic arena[37]. In addition, other issues such as motivations for partner abuse because of nondomestic factors such as women's financial status, employment position, education and husband's drunkenness are not included in the measure of attitudes toward IPVAW. Apart from instrumental validity, the potential limitations of face-to-face interviews need to be acknowledged[37]. For example, when contrasted with self administered questionnaires, participants may tend to underreport their attitudes toward IPVAW in the presence of their interviewers. However, ethical measures such as guarantees of anonymity and administering the interviews by trained personal may have improved such reporting[37].

Despite these limitations, the study strengths are significant. It is a large, population-based study with national coverage. In addition, data of the DHS are widely perceived to be of high quality, as they were based on sound sampling methodology with high response rate. DHS also adhere to stringent ethical rules in the collection of domestic violence data used. An important strength of this study is the number of included countries and geographic and socioeconomic diversities constitute a good yardstick for the region, and help to strengthen the findings from the study.

Conclusion

This large comparative analysis has provided evidence that IPVAW was widely acceptable under certain circumstances and more such among women, younger people, less educated, poorest, those living in rural areas, those with less access to media and single decision makers. There is a need for proactive efforts to break the norms that sustain women's vulnerability in the society besides socio-economic development as well as promotion of higher education among men and women. Direct concerted efforts from the government, non-governmental organisations and enlightened men and women within the society are necessary to raise awareness about the issue as well as questioning the social norms. This study has provided information about individual predictors of attitudes toward IPVAW in 17 sub-Saharan countries. However, our knowledge about the contextual factors associated with the attitudes toward IPVAW is still limited.

Competing Interests

The authors declare that they have no competing interests.

Authors' Contributions

OAU, SL and TM were involved in the conception of the study. OAU carried out data extraction. OAU conducted statistical analysis under supervision of SL and TM. OAU drafted the paper with contributions from the co-authors. All authors read and approved the final manuscript.

Acknowledgements

The data used in this study were made available through MEASURE DHS Archive. The data were originally collected by the ORC Macro, Calverton USA. Neither the original collectors of the data nor the Data Archive bear any responsibility for the analyses or interpretations presented in this project. Without wishing to implicate them in any way, the authors are grateful to Laura McCloskey,

Lori Post, Sarah Cook, and Jhumka Gupta for critical review of an earlier version of this manuscript.

References

1. Ofei-aboagye RO: Domestic violence in Ghana: an initial step. Columbia J Gend Law 1994, 4(1):1–25.

2. Okemgbo CN, Omideyi AK, Odimegwu CO: Prevalence, patterns and correlates of domestic violence in selected Igbo communities of Imo State, Nigeria. African journal of reproductive health 2002, 6(2):101–114.

3. Kiragu J: Policy review: HIV prevention and women's rights: working for one means working for both. AIDS Captions 1995, 2(3):40–46.

4. Rivera Izabal LM: Women's legal knowledge: a case study of Mexican urban dwellers. Gend Dev 1995, 3(2):43–48.

5. Oyediran KA, Isiugo-Abanihe U: Perceptions of Nigerian women on domestic violence: evidence from 2003 Nigeria Demographic and Health Survey. African journal of reproductive health 2005, 9(2):38–53.

6. Watts C, Ndlovu M, Njovana E, Keogh E: Women, violence and HIV/AIDS in Zimbabwe. Safaids News 1997, 5(2):2–6.

7. Campbell J, Jones AS, Dienemann J, Kub J, Schollenberger J, O'Campo P, Gielen AC, Wynne C: Intimate partner violence and physical health consequences. Arch Intern Med 2002, 162(10):1157–1163.

8. Moore M: Reproductive health and intimate partner violence. Fam Plann Perspect 1999, 31(6):302–306.

9. Moore TM, Stuart GL, Meehan JC, Rhatigan DL, Hellmuth JC, Keen SM: Drug abuse and aggression between intimate partners: a meta-analytic review. Clin Psychol Rev 2008, 28(2):247–274.

10. Golding JM: Intimate partner violence as a risk factor for mental disorders a meta-analysis. Journal of Family Violence 1999, 14(2):99–132.

11. Ellsberg M, Jansen HA, Heise L, Watts CH, Garcia-Moreno C: Intimate partner violence and women's physical and mental health in the WHO multi-country study on women's health and domestic violence: an observational study. Lancet 2008, 371(9619):1165–1172.

12. Emenike E, Lawoko S, Dalal K: Intimate partner violence and reproductive health of women in Kenya. Int Nurs Rev 2008, 55(1):97–102.

13. Lawoko S, Dalal K, Jiayou L, Jansson B: Social inequalities in intimate partner violence: a study of women in Kenya. Violence Vict 2007, 22(6):773–784.

14. Owoaje ET, Olaolorun FM: Intimate partner violence among women in a migrant community in southwest Nigeria. Int Q Community Health Educ 2005, 25(4):337–349.

15. Karamagi CA, Tumwine JK, Tylleskar T, Heggenhougen K: Intimate partner violence against women in eastern Uganda: implications for HIV prevention. BMC Public Health 2006, 6:284.

16. Abrahams N, Jewkes R, Laubscher R, Hoffman M: Intimate partner violence: prevalence and risk factors for men in Cape Town, South Africa. Violence Vict 2006, 21(2):247–264.

17. McCloskey LA, Williams C, Larsen U: Gender inequality and intimate partner violence among women in Moshi, Tanzania. Int Fam Plan Perspect 2005, 31(3):124–130.

18. Watts C, Mayhew S: Reproductive health services and intimate partner violence: shaping a pragmatic response in Sub-Saharan Africa. Int Fam Plan Perspect 2004, 30(4):207–213.

19. Fawole OI, Aderonmu AL, Fawole AO: Intimate partner abuse: wife beating among civil servants in Ibadan, Nigeria. African journal of reproductive health 2005, 9(2):54–64.

20. Choi SY, Ting KF: Wife beating in South Africa: an imbalance theory of resources and power. J Interpers Violence 2008, 23(6):834–852.

21. Deyessa N, Kassaye M, Demeke B, Taffa N: Magnitude, type and outcomes of physical violence against married women in Butajira, southern Ethiopia. Ethiop Med J 1998, 36:83–5.

22. Silverman JG, Gupta J, Decker MR, Kapur N, Raj A: Intimate partner violence and unwanted pregnancy, miscarriage, induced abortion, and stillbirth among a national sample of Bangladeshi women. BJOG 2007, 114(10):1246–1252.

23. Gupta J, Silverman JG, Hemenway D, Acevedo-Garcia D, Stein DJ, Williams DR: Physical violence against intimate partners and related exposures to violence among South African men. CMAJ 2008, 179(6):535–541.

24. Johnson KB, Das MB: Spousal violence in bangladesh as reported by men: prevalence and risk factors. J Interpers Violence 2009, 24(6):977–995.

25. Silverman JG, Decker MR, Kapur NA, Gupta J, Raj A: Violence against wives, sexual risk and sexually transmitted infection among Bangladeshi men. Sex Transm Infect 2007, 83(3):211–215.

26. Visaria L: Violence against women: a field study. Econ Pol Wkly 2000, 1742–1751.

27. Jewkes R: Intimate partner violence: causes and prevention. Lancet 2002, 359(9315):1423–1429.

28. Faramarzi M, Esmailzadeh S, Mosavi S: A comparison of abused and non-abused women's definitions of domestic violence and attitudes to acceptance of male dominance. European journal of obstetrics, gynecology, and reproductive biology 2005, 122(2):225–231.

29. Hanson RK, Cadsky O, Harris A, Lalonde C: Correlates of battering among 997 men: family history, adjustment, and attitudinal differences. Violence Vict 1997, 12(3):191–208.

30. Gage AJ, Hutchinson PL: Power, control, and intimate partner sexual violence in Haiti. Arch Sex Behav 2006, 35(1):11–24.

31. Gracia E: Unreported cases of domestic violence against women: towards an epidemiology of social silence, tolerance, and inhibition. J Epidemiol Community Health 2004, 58(7):536–537.

32. Biden JR Jr: Violence against women. The congressional response. Am Psychol 1993, 48(10):1059–1061.

33. Gracia E, Herrero J: Acceptability of domestic violence against women in the European Union: a multilevel analysis. J Epidemiol Community Health 2006, 60(2):123–129.

34. Uthman OA, Moradi T, Lawoko S: The independent contribution of individual-, neighbourhood-, and country-level socioeconomic position on attitudes towards intimate partner violence against women in sub-Saharan Africa: a multilevel model of direct and moderating effects. Soc Sci Med 2009, 68(10):1801–1809.

35. Rani M, Bonu S, Diop-Sidibe N: An empirical investigation of attitudes towards wife-beating among men and women in seven sub-Saharan African countries. African journal of reproductive health 2004, 8(3):116–136.

36. Lawoko S: Factors associated with attitudes toward intimate partner violence: a study of women in Zambia. Violence Vict 2006, 21(5):645–656.

37. Lawoko S: Predictors of attitudes toward intimate partner violence: a comparative study of men in Zambia and Kenya. J Interpers Violence 2008, 23(8):1056–1074.

38. Hindin MJ: Understanding women's attitudes towards wife beating in Zimbabwe. Bull World Health Organ 2003, 81(7):501–508.

39. Publications by country, [http://www.measuredhs.com/pubs/browse_region.cfm].

40. Vyas S, Kumaranayake L: Constructing socio-economic status indices: how to use principal components analysis. Health Policy Plan 2006, 21(6):459-468.

41. Filmer D, Pritchett LH: Estimating wealth effects without expenditure data – or tears: an application to educational enrollments in states of India. Demography 2001, 38(1):115–132.

42. Montgomery MR, Gragnolati M, Burke KA, Paredes E: Measuring living standards with proxy variables. Demography 2000, 37(2):155–174.

43. DerSimonian R, Laird N: Meta-analysis in clinical trials. Controlled clinical trials 1986, 7(3):177–188.

44. Cochran WG: The combination of estimates from different experiments. Biometrics 1954, 8:101–129.

45. Higgins JP, Thompson SG: Quantifying heterogeneity in a meta-analysis. Stat Med 2002, 21(11):1539–1558.

46. Ender P: Applied Categorical & Nonnormal Data Analysis: Collinearity Issues. UCLA: Academic Technology Services, Statistical Consulting Group. [http://www.gseis.ucla.edu/courses/ed231c/notes2/collin.html]

47. Hocking RR: Methods and Applications of Linear Models. New York: Wiley; 1996.

48. Gracia E, Herrero J: Perceived neighborhood social disorder and attitudes toward reporting domestic violence against women. J Interpers Violence 2007, 22(6):737–752.

49. Sabol WJ, Coulton CJ, Korbin JE: Building community capacity for violence prevention. J Interpers Violence 2004, 19(3):322–340.

50. Rani M, Bonu S: Attitudes toward wife beating: a cross-country study in Asia. J Interpers Violence 2009, 24(8):1371–1397.

Psychological Wellbeing, Physical Impairments and Rural Aging in a Developing Country Setting

Melanie A. Abas, Sureeporn Punpuing,
Tawanchai Jirapramupitak, Kanchana Tangchonlatip
and Morven Leese

ABSTRACT

Background

There has been very little research on wellbeing, physical impairments and disability in older people in developing countries.

Methods

A community survey of 1147 older parents, one per household, aged sixty and over in rural Thailand. We used the Burvill scale of physical impairment, the

Thai Psychological Wellbeing Scale and the brief WHO Disability Assessment Schedule. We rated received and perceived social support separately from children and from others and rated support to children. We used weighted analyses to take account of the sampling design.

Results

Impairments due to arthritis, pain, paralysis, vision, stomach problems or breathing were all associated with lower wellbeing. After adjusting for disability, only impairment due to paralysis was independently associated with lowered wellbeing. The effect of having two or more impairments compared to none was associated with lowered wellbeing after adjusting for demographic factors and social support (adjusted difference -2.37 on the well-being scale with SD = 7.9, p < 0.001) but after adjusting for disability the coefficient fell and was non-significant. The parsimonious model for wellbeing included age, wealth, social support, disability and impairment due to paralysis (the effect of paralysis was -2.97, p = 0.001). In this Thai setting, received support from children and from others and perceived good support from and to children were all independently associated with greater wellbeing whereas actual support to children was associated with lower wellbeing. Low received support from children interacted with paralysis in being especially associated with low wellbeing.

Conclusion

In this Thai setting, as found in western settings, most of the association between physical impairments and lower wellbeing is explained by disability. Disability is potentially mediating the association between impairment and low wellbeing. Received support may buffer the impact of some impairments on wellbeing in this setting. Giving actual support to children is associated with less wellbeing unless the support being given to children is perceived as good, perhaps reflecting parental obligation to support adult children in need. Improving community disability services for older people and optimizing received social support will be vital in rural areas in developing countries.

Background

There is increasing interest worldwide in the study of wellbeing as a means to assess need and to evaluate positive dimensions of health care programs. Positive mental health "which allows individuals to realise their abilities, cope, and contribute to their communities" [1] and the capacity to sustain social relationships are key dimensions of wellbeing [2]. Wellbeing can be measured in terms of positive psychological symptoms (such as being able to enjoy things and to let go of worries) or life satisfaction, but increasingly multidimensional scales are used which include concepts such as autonomy, self-acceptance and relations with others [3,4].

Research on associations between physical impairments and wellbeing in older people has been limited [5-7] although there have been several studies of depression as an outcome suggesting that disability mediates most of the effect of specific medical conditions on depression [8-10]. However, research until now has come almost entirely from richer industrialised countries. One aim of this study was to see whether patterns of association between impairment, disability and psychological well-being in Thailand are similar to or different from those described elsewhere. Given cross-cultural differences in perceived well-being, a recent advance has been to develop culture-specific scales such as the Chinese Aging Well Profile (2007) [11]. In Thailand, Ingersoll-Dayton et al [12] developed and validated the Thai psychological well-being scale, which is related to the Scale of Psychological Well-being Scale [3]. Particular features of this, which is the only multidimensional wellbeing scale developed for use with Thai older people, is that compared to versions used in Western settings, more of the dimensions are interpersonal (measuring harmony and interconnectedness with other people) and fewer are intrapersonal (e.g. measuring acceptance and positive mood).

In Thailand, the setting for this study, the proportion of adults 60 years of age and over rose from 4.5% in 1960 to 9.5% in 2000 and is predicted to be 25% in 2040[13]. In the rural Thai context, as in many developing countries, facilities for health care and support for disabilities are limited. Also in many other developing countries, rapid rise in rural to urban migration of young adults means that older parents are increasingly living separately from their adult children [14]. In Thailand as in other Asian cultures, children traditionally take responsibility for older parents and older parents continue to support children. Given the potential relative importance of support from children [15] we were interested to see if support from children rather than support from others was associated with wellbeing.

Methods

Setting

We nested the study within the Kanchanaburi Demographic Surveillance System in western Thailand [16]. Kanchanaburi province is a mostly rural region located 130 kilometres west of Bangkok with a population of about 735,000 in 2007. The Kanchanaburi Demographic Surveillance System system has monitored households since 2000 in 100 neighborhoods (villages and urban census blocks). The neighborhoods were drawn from five strata (classified on ecological, socio-economic and population criteria) by stratified random sampling from the province population of 871 villages and 131 urban census blocks. The study described here is part of a longitudinal study designed to study the impact on older parents of out-migration of their adult children/offspring[17] During sampling for the main study we needed to identify which older adults were parents of at least one living child offspring, and whether the older parent was co-resident or not with at least one of their offspring. There was a potential sample of 3916 households with at least one older adult aged 60 and above, of whom 2432 (62%) had at least one child offspring of the older adult in the same household, and 1484 (38%) did not. We used simple random sampling to select 60% of households where an older adult was not co-resident with at least one of their child offspring and 30% of households where an older adult was co-resident with at least one of their child offspring. This comprised a total of 1620 households. We used random selection to identify the participant in situations where there was more than one eligible parent living in a household. Data were collected from November 2006 to Jan 2007.

Recruitment

The interviewing team visited each sampling unit and made contact with the village headman prior to visiting each selected household. The populations were mostly already well acquainted with the demographic surveillance system. If the selected older adult and the household head gave consent, the interviewer first interviewed the household head with the household questionnaire and then the older adult with the individual questionnaire.

Questionnaire Development

We carried out focus group discussions to explore experiences of rural ageing, health and wellbeing and exchanges with family members. This informed the development of the questionnaire which was pre-tested by a team of ten experienced interviewers on three separate occasions. After each pre-test we made modifications by consensus. The final version was back-translated to English and checked for consistency by a bilingual psychiatrist and a bilingual social scientist.

Inclusion Criteria

Fluent Thai-speaking; aged 60 or over; parent of at least one living child (biological, adopted or step-child); residence in a demographic surveillance system village since at least 2004.

Dependent Variable

Psychological well-being. We used the 15-item Thai well-being scale [12,18], developed using extensive qualitative and quantitative methods. It has five dimensions of wellbeing which are harmony, interdependence with close persons, respect (from others), acceptance and enjoyment. Each dimension has three items which were developed from confirmatory factor analysis. We used the global factor model which was shown in Thailand to have good fit indices (goodness of fit 0.95, root mean square error of approximation 0.05) [12]. The items of the scale have been shown to have adequate internal consistency (Cronbach's alpha coefficient in this sample 0.89) and test-retest reliability (ranging from 0.6 to 0.7 in previous work) [12] and the scale correlated positively with life satisfaction and negatively with the Geriatric Depression Scale (-0.4) [12]. A statement is read out for each item. For example, for acceptance the statement is 'When you have small problems, you can let go of your worries.' The older person indicates on a 4-point scale if the statement is not at all true, slightly true, somewhat true or very true.

Independent Variable

Physical Illnesses and Impairments: we used a modified version of the Burvill physical illness scale [19]. Participants were asked about the presence of 13 common medical problems including breathlessness, faints/blackouts, arthritis, paralysis/loss of limb, skin disorders, hearing difficulties, heart trouble, eyesight problems, gastrointestinal problems, high blood pressure, diabetes and pain. If any of the problems was present we rated it as impairment if participants stated that the problem was interfering a great deal with their function.

Potential Confounders

Socio-Economic Position

Years of education, number of household assets (out of 22, such as ownership of a fridge, motorcycle, or mobile phone), and household wealth index. We used principal components analysis to develop the household wealth index from the list of assets and the interviewer's global rating of household quality. The first principal component (which accounted for 26% of the variance compared to 7% for the second next most important) was used to provide an overall socioeconomic index based on these 23 items. This final index comprised 15 items (14 household assets plus household quality).

Social Network and Social Support

We modified existing measures in the light of the importance in the Thai context of the family and of children. We measured size of neighbourhood family network, frequency of talking to a child, frequency of talking to friends, received support (instrumental, emotional, financial), actual support to children (instrumental, emotional, financial), perceived adequacy of support from and to children, and received support from others [20-22]. The received social support from children scale rated received support yes/no from any of their children on each of ten items. The received social support from others scale rated received support yes/no from anyone other than children on the same ten items. The support to children scale rated support to any children on each of five items.

Cognitive Function

We used a learning task which has been used extensively in low and middle income countries which is drawn from the Consortium to Establish a Registry of Alzheimer's Disease (CERAD) [23,24], comprising immediate recall and delayed recall of a ten-word list. We defined significant cognitive impairment as performance at or below 1.5 standard deviations below the norm for the individual's age group and educational level on both tests.

Disability

We used the brief (12-item) questionnaire from the WHO Disability Assessment Schedule to rate disability over the past 30 days [25]. We were unable to translate the item on learning a new task, which was viewed as not applicable for older adults in this setting. Therefore, we used 11 items, each self-rated on a four point scale from no problem with carrying out the activity to total/extreme inability. Domains included understanding and communicating with the world, getting around, self-care, getting along with people, activities and participation in society. We categorised the total score into thirds of low, medium and high disability.

Data Collection

The data collection team of four supervisors and twelve interviewers had at least a bachelor's degree. Most had previous experience with interviewing for the demographic surveillance system. Residential training took ten days and included presentations, role play and practice in pilot villages. The study was presented to the interviewers as a study of healthy ageing in Thailand. Purposefully, no possible links were discussed between psychological wellbeing, impairment, disability or social support from children in order to blind the interviewers to the research hypotheses and none of these sections of the interview immediately followed each other in sequence.

The data collection team stayed in the villages at the headman's house or the temple. Quality control included checks on data completeness and consistency. Interviewers had to return to the participant if data were inadequate. Field station research managers (trained in the interview but blind to the hypothesis), and researchers were in frequent telephone contact and regularly visited the data collection teams. We conducted all interviews in Thai and gathered informed consent from all participants. We gained ethical approval from Kings College Research Ethics Committee (No. 05/05-68) and from Mahidol University Institutional Review Board.

Sample Size Calculation

This was developed for the main longitudinal study which was designed to study the impact on older parents of out-migration of their adult children/offspring from the district [17]. The sample size was based on a comparison of prevalence of common mental disorder in those with all children migrated versus those with some children migrated and required a total sample size of 954 given the proportions expected of those exposed and not exposed to having all their children migrate from the district.

Analysis

We used Stata version 9 for Windows (Release 9, College Station, TX: Stata Corporation. 2003). We weighted the data using the product of two sets of probability weights to take account of differential sampling at neighbourhood and household levels. The weighting at neighbourhood level took account of the probability of the neighbourhood being selected from the total number of neighbourhoods in that stratum in the province. The weighting at household level took account of the probability of being selected if the older parent was or was not co-resident with one of their offspring. We used the survey commands in Stata (svyset) for analyses. We first described the unadjusted associations between wellbeing score and the socio-economic, social support and health variables. We modelled impairment in two ways: as individual impairments and as a total of different impairments (one impairment versus none and two or more versus none). We used multiple linear regressions to develop a model for the effect of impairment on wellbeing, carrying out tests of the effect of impairment after adding in potential confounding variables. We explored interactions between social support, specific impairments, total impairments and total disability in the multivariable model. All tests were Wald tests as appropriate for weighted survey data. Residuals were computed for the final multivariable model and plotted as histograms (to assess any evidence for non normality, including individual outliers) and were also plotted against predicted values (to assess evidence for heteroscedasicity, in the sense of greater spread with increasing value). Variance inflation factors (VIFs) were computed for all independent variables to check for collinearity.

Results

1620 older adults in 1620 households were sampled, of whom 1300 (80%) were eligible to take part. Reasons for not being eligible were having no biological or adopted children or step-children; having died since 2004, or moved out of the village. Out the 1300 eligible, 1147 (88%) agreed to take part and 153 (12%) were non-responders of whom 110 were unavailable for an interview (despite at least three visits to the household), 21 refused to take part and 22 were too unwell. Of the responders, data were incomplete for 43 due to the older adult being unwell or cognitively impaired. There were no significant differences between responders and non-responders in terms of age, gender, living alone, being married, or education.

Table 1. Descriptive characteristics of parents: actual sample numbers (total n = 1147) and weighted percentages

	Study sample n = 1147	Weighted percentages
Female	n = 634	57%
Working	n = 564	48%
Marital status:		
Married	n = 633	54%
Widowed	n = 451	41%
Divorced/separated/single	n = 63	6%
Live alone	n = 155	9%
Education:		
None	n = 332	28%
1–3 years	n = 174	15%
Primary (4 yrs)	n = 541	49%
More than primary	n = 99	8%
Proportion with two or more limiting physical impairments	n = 540	50%
Cognitive impairment	n = 91	8%
At least one child living at home	n = 551	63%

Table 1 shows the actual sample numbers and weighted estimate of the characteristics in the wider province population of parents from which the sample was drawn. The average age was 70 years (SD 7.1). As shown in Table 1, 57% of the participants were female. Nearly half had less than primary school education, which for our sample meant less than four years education. (Only in the last two decades has Thailand's compulsory education extended to six and now to twelve years) Nearly half were still working. Because we over-sampled those not co-resident with a child, the study population has a lower proportion living with a child compared to the province estimate and is slightly more likely to live alone. Otherwise there were negligible differences between the study sample and the estimated province population. The average number of live children in these parents was 4.8 (SD 2.4); 2.4 sons and 2.4 daughters. Three-quarters either lived with a child or saw a child daily. The mean duration of residence in the same district was nearly 50 years. The mean wellbeing score was 33.3 (SD 7.6).

Table 2. Prevalence of impairments and associations with wellbeing, weighted linear regression

Health impairments	Weighted percentages (95% confidence intervals)	Coefficient for association with wellbeing	P value for association with wellbeing	P value for association with wellbeing, adjusted for disability
Arthritis or rheumatism	44.4 (40.0–48.4)	-1.66	<0.001	0.915
Eyesight	23.3 (19.3–27.3)	-2.07	<0.001	0.202
Hearing	7.6 (6.0–9.2)	-.76	0.496	0.843
Cough	3.9 (2.4–5.4)	-2.89	0.110	0.306
Breathing	7.7 (5.4–10.0)	-2.73	0.024	0.186
High blood pressure	16.3 (13.0–19.5)	-0.48	0.415	0.185
Diabetes	7.1 (4.8–8.7)	-1.57	0.263	0.788
Heart trouble or angina	6.4 (4.1–8.7)	-1.12	0.534	0.831
Stomach or intestine	9.3 (6.6–12.0)	-2.50	0.008	0.086
Faints or blackouts	17.8 (14.5–20.9)	-2.63	0.001	0.143
Paralysis	2.3 (1.1–3.5)	-4.66	<0.001	0.012
Skin	3.4 (2.2–4.6)	0.02	0.993	0.785
Pain	37.3 (32.1–42.4)	-2.46	<0.001	0.105

The three most common impairments were arthritis, pain, and eyesight problems. Approximately one-third (32%) of the older adults did not have any impairment, 18% had one and 50% had two or more impairments. Impairments due to arthritis, pain, paralysis, vision, stomach problems or breathing were all associated with lowered wellbeing. Paralysis, faints/blackout, breathlessness, and pain were the impairments with the highest effect size for less wellbeing. After adjusting the impairments for disability, only paralysis remained significantly associated with low wellbeing.

Table 3. Association between wellbeing score and having one or two or more physical impairments (sample n = 1147)

Number of physical impairments	Coefficient for having one impairment compared to none *	Coefficient for having two or more impairments compared to none *	Wald test F(2, 95)	P value
	-1.55	-3.03	15.52	<0.001
Adjusted for socio-demographic characteristics[1]	-1.01	-2.55	8.52	<0.001
Adjusted for 1 + social support and social network[2]	-0.64	-2.43	13.44	<0.001
Adjusted for 1 + 2 + social support to children[3]	-0.53	-2.37	14.42	<0.001
Adjusted for 1 + 2 + 3 + disability[4]	-0.23	-0.48	0.42	0.656
Adjusted for 1 + 2 + 3 + 4 + cognitive impairment[5]	-0.21	-0.46	0.38	0.685

As shown in Table 3, having one impairment compared to none and having two or more compared to none was significantly associated with less wellbeing. This association remained after adjusting for socio-demographic factors, social support from children, social support to children, and social support from others. There appeared to be some positive confounding by socio-demographic factors as the coefficients for the association with impairment fell slightly and the statistical significance decreased. This may be explained because factors such as wealth and education are associated with greater wellbeing and with less impairment. There appeared to be some slight negative confounding by social support from and to children as the significance rose again after adjusting for these. This could be because more impaired older peoples are likely to receive more social support from children and others, and more social support is also associated with greater wellbeing. Finally, after adjusting for disability, the association between number of impairments and wellbeing fell and was no longer significant.

Table 4. Associations between psychological wellbeing and demographic, social and physical health status (sample n = 1147)

	Unadjusted Coefficient	Unadjusted P value	Adjusted coefficient[a]	Adjusted P value[a]
Older Age (years)	0.03	0.455	0.13	0.010
Female	-1.25	0.027	0.02	0.980
Currently working	0.43	0.419	1.32	0.018
Married versus widowed/single/divorced	1.05	0.066	0.33	0.581
Live alone	-1.06	0.113	-0.36	0.659
Education	1.48	0.007	0.15	0.145
Wealthy household	0.89	<0.001	0.32	<0.001
Physical impairment	-0.75	<0.001	-.37	0.020
Paralysis	-4.66	<0.001	-2.96	<0.001
Disability	-0.29	<0.001	-.22	<0.001
Cognitive impairment	-1.03	0.218	-1.43	0.223
Family social network size	0.11	<0.001	0.07	0.002
At least one child living in household versus no children in the household	-0.10	0.850	-0.57	0.304
Talk to a child at least weekly	0.93	0.002	0.74	0.029
Receiving support from children	0.51	<0.001	3.06	<0.001
Receiving financial remittances from children	2.18	<0.001	1.55	<0.001
Giving support to children	0.25	0.284	-0.62	<0.001
Receiving support from others	0.44	0.003	0.53	0.003
Perceive good support from children	3.79	<0.001	3.06	<0.001
Perceive giving good support to children	3.31	<0.001	1.26	0.029

[a] adjusted for all other variables in the table in a weighted regression.

Variables that were significantly associated with wellbeing either before and/or after adjustment are shown in Table 4. The parsimonious multivariable model for psychological wellbeing included age, household wealth, currently working, family network size close-by, receiving support from children, receiving support from others, talking more frequently to a child, perceiving receiving very adequate support from children, perceiving giving good support to children, less impairment due to paralysis, (p = 0.003), less general impairment, less disability, and giving less actual support to children. Of note, neither living alone or cognitive impairment were associated with wellbeing. The percentage of variance explained by the multivariable model was 32%. The residuals showed no evidence for non normality nor for outliers, and there was no evidence for heteroscedascity. There was no evidence for collinearity (all VIFs <10).

There was an interaction between social support from children and paralysis – those with low received social support from children and with paralysis were especially likely to have low wellbeing (p value for interaction 0.033).

Discussion

The key finding from this paper is that impairment due to paralysis was associated with lowered psychological wellbeing in older Thai people, even after controlling for eleven other physical impairments, disability, socio-economic factors and social support. A second key finding is that while an increasing number of impairments was also associated with less wellbeing, this association, and those with other individual impairments, were explained by disability. A third finding is that in this Thai setting, received support from adult child offspring, received support from others and perceived support from adult child offspring were all independently associated with greater wellbeing in older parents whereas actual support to children was associated with lower wellbeing.

Chance is an unlikely explanation for the adjusted association between paralysis and low wellbeing, and for the adjusted association between disability and low wellbeing, as the associations were significant at a level of p = 0.001. We were able to adjust for a range of covariates so confounding is an unlikely explanation. All impairment and disability measures relied on subjective perception which may lead to misclassification of health status, although a high level of agreement has been reported between self-reported and objective health status measures [19]. Bias is unlikely in this community sample with a good response rate and interviewers were blind to the study hypotheses. Although we oversampled, this was on the basis on living arrangements rather than health and was anyway taken account of in the analysis. Non-systematic error is possible – for instance this might have come about through poor reliability of the interviewing team or through participants' errors in recall of their health problems, although previous work has shown a high level of agreement between self-report and objective health status measures [19]. We did not formally assess inter-rater reliability. However as part of the demographic surveillance system approach, quality control is well established and prioritised including daily checks on data completeness and consistency, having a research supervisor for each team of interviewers and having field station research managers (trained in the interview but blind to the hypothesis), and researchers, in frequent telephone contact and making regular visits to the data collection teams. This is a cross-sectional study so the direction of causality cannot be definitely inferred.

Why was paralysis associated with a large and significant effect on wellbeing? Studies of older people in Western countries have reported low mood and depression particularly following stroke and that this association was independent of disability [26]. Post-stroke depression of course may have a biological basis which may explain our finding [27]. However, wellbeing is a broader concept than depression. Our measure of wellbeing was developed and validated using thorough qualitative and quantitative work with Thai older people [12,18] and includes concepts vital to Thai wellbeing including interpersonal as well as intrapersonal aspects. The effect of paralysis may be due to the scarce disability services in rural Thailand, with few opportunities to receive aids, adaptations, or community transport. Rural people may thus be especially vulnerable to loss of social contacts in the neighbourhood and to losing respect. Another possibility is that impacts of stroke go beyond disability, either via biological effects on the brain [27] or through the psychological meaning of stroke such as shame over loss of function and altered appearance and fears about prognosis. In this setting of high out-migration, absence of children may also be a factor, although most older people still either live close to a child or talk to a child weekly or more.

Our finding that disability explains the association between number of impairments and low wellbeing echoes studies that have looked at impairment, disability and depression and at impairments and wellbeing in Western countries [6,9,28,29]. Prospective studies have shown that disability can predict the onset of depression [29]. A recent review concluded that much of the effect of impairment on negative affect could be explained by the potential mediating effect of disability [30]. It is striking that our result mirrors that from western countries, showing the cross-cultural applicability of the wellbeing model.

The model for greater wellbeing included other factors, notably received social support from children, perceived social support from children, received social support from others, financial remittances from children and wealth. As a number of associations were analysed in this study, a problem of multiple testing might have occurred. However, it is unlikely that this would explain our findings as most of the factors in the parsimonious model for wellbeing were significant at $p < 0.001$ or $p = 0.001$. Several possible mechanisms could explain the effect of received social support on wellbeing. Social support may reduce stress and consequently buffer the effect of negative events. Although received support is likely to reflect need, certain types of received support may be valuable in bringing about improved wellbeing[31].

Greater social support might also aid older people with impairment to carry out daily tasks, encourage them to be physically active, increase medication compliance, decrease social restriction and enhance self-esteem [32]. In the Thai culture, connections between parents and children are vital [33]. Although many parents in this study had out-migrant children, they continued to receive support through telephone contact, visits and economic remittances[17] In addition, they received support from others, often neighbours or other relatives living close by, and this was also independently associated with greater wellbeing. This suggests that older people living without children are adapting to the realities of out-migration and finding help from others close by in their neighbourhood. It is striking that received support from children and from others appeared helpful, and that received support from children may even buffer the impact of paralysis on low wellbeing. Older Thai people may place less value on autonomy than those in western countries, finding support from family members especially important and comforting [12]. A perception by the parent of giving a good amount of support to their offspring was associated with better well-being. However, giving actual support to children was associated with less wellbeing, perhaps reflecting parental obligation in this culture to support adult children in need [34].

Some limitations of this study include its cross-sectional design. Secondly our measure of wellbeing is culture specific—although this may also be regarded as strength of the study. Thirdly, the findings from this study might lack generalisability to all older adults as the sample was restricted to parents with at least one living child, although in Thailand this excluded only 5% of older people as we included anyone with a biological, adopted or stepchild.

In conclusion, disability may mediate most of the impact of chronic physical impairments on psychological wellbeing, although paralysis appears to have an independent effect. Received social support, perceived social support and wealth also have important positive effects on psychological wellbeing. Improving disability services and optimising social support will be vital in rural areas in developing countries which are likely to experience increasing depletion of younger adults in the next decade. While care is currently provided by family members, especially daughters and grand-daughters, we suggest that potentially valuable services in rural areas may include home care programmes for older people and their carers, home visits by health care volunteers in the village, day care, extending the existing network of 'elderly clubs,' occupational therapy to enable aids and adaptations at home, and making a range of facilities more accessible to older disabled people.

Conclusion

In conclusion, in this Thai rural setting, most of the association between physical impairments and lower wellbeing in older people is explained by disability.

Received support from children and from others and perceived high support from and to children were all independently associated with greater wellbeing whereas giving actual support to children was associated with lower wellbeing. Improving community disability services for older people and optimizing received social support through families, neighbours and home care programs will be vital in rural areas in developing countries.

Competing Interests

The authors declare that they have no competing interests.

Authors' Contributions

All authors made substantial contributions to study design and interpretation of data. MA had main responsibility for analysing data and drafting the manuscript. SP and KT had main responsibility for acquisition of data. All authors were involved in revising the manuscript critically and have given final approval of the version to be published.

Acknowledgements

We thank Dr Bencha Yoddumnern-Attig, Dr Philip Guest and Prof Martin Prince for advice on the study design and methods, Ms Wannee Hutapat and Ms Jongjit Rithirong for data management, Dr Robert Stewart for comments on the manuscript, all the field staff (Niphon Darawuttimaprakorn, Jeerawan Hongthong, Phattharaphon Luddakul Wipaporn Jarruruengpaisan and Yaowalak Jiaranai) and participants of the Kanchanaburi Demographic Surveillance System, and the Wellcome Trust for funding the project (WT 078567).

References

1. World Health Organisation: Promoting Mental Health; Concepts emerging evidence and practise, Summary report.Geneva. 2004.

2. World Health Organisation: Strengthening mental health promotion. Geneva. 2001.

3. Ryff CD, Keyes CLM: The structure of psychological well-being revisited. Journal of Personality and Social Psychology 1995, 69(4):719–727.

4. Tennant R, Hiller L, Fishwick R, Platt S, Joseph S, Weich S, Parkinson J, Secker J, Stewart-Brown S: The Warwick-Edinburgh Mental Well-being Scale (WEMWBS): development and UK validation. Health Qual Life Outcomes 2007, 5:63.

5. Araki A, Murotani Y, Kamimiya F, Ito H: Low Well-Being Is an Independent Predictor for Stroke in Elderly Patients with Diabetes Mellitus. Journal of the American Geriatrics Society 2004, 52(2):205–210.

6. Kendig H, Browning CJ, Young AE: Impacts of illness and disability on the well-being of older people. Disabil Rehabil. 2000, 22(1–2):15–22.

7. Smith J: Well-being and health from age 70 to 100: findings from the Berlin Aging Study. European Review 2001, 9(04):461–477.

8. Beekman AT, Penninx BW, Deeg DJ, Ormel J, Braam AW, van Tilburg W: Depression and physical health in later life: results from the Longitudinal Aging Study Amsterdam (LASA). J Affect Disord 1997, 46(3):219–231.

9. Ormel J, Kempen GI, Penninx BW, Brilman EI, Beekman AT, van Sonderen E: Chronic medical conditions and mental health in older people: disability and psychosocial resources mediate specific mental health effects. Psychol Med 1997, 27(5):1065–1077.

10. Prince MJ, Harwood RH, Blizard RA, Thomas A, Mann AH: Social support deficits, loneliness and life events as risk factors for depression in old age. The Gospel Oak Project VI. Psychol Med 1997, 27(2):323–332.

11. Ku PW, Fox K, McKenna J: Assessing Subjective Well-being in Chinese Older Adults: The Chinese Aging Well Profile. Social Indicators Research 2008, 87(3):445–460.

12. Ingersoll-Dayton B, Saengtienchai C, Kespichayawattana J, Aungsuroch Y: Measuring psychological well-being: insights from Thai elders. Gerontologist 2004, 44(5):596–604.

13. Skeldon R: Migration of Women in the Context of Globalization in the Asian and Pacific Region. [http://www.unescap.org/esid/GAD/Publication/DiscussionPapers/02/series2.pdf]. Women in Development Discussion Paper Series ESCAP; 1999.

14. Deshingkar P: Internal migration, poverty and development in Asia. [http://www.odi.org.uk/resources/download/29.pdf]. Overseas Development Institute; 2006.

15. Chou KL, Chi I: Reciprocal relationship between social support and depressive symptoms among Chinese elderly. Aging & Mental Health 2003, 7(3):224–231.

16. Report on baseline survey round 1 Nakhon Pathom: Institute for Population and Social Research, Mahidol University, 2002; 2000.

17. Abas MA, Punpuing S, Jirapramukpitak T, Guest P, Tangchonlatip K, Leese M, Prince M: Rural-urban migration and depression in ageing family members left behind. British Journal of Psychiatry 2009, 195:54–60.

18. Ingersoll-Dayton B, Saengtienchai C, Kespichayawattana J, Aungsuroch Y: Psychological well-being Asian style: the perspective of Thai elders. J Cross Cult Gerontol 2001, 16(3):283–302.

19. Burvill PW, Mowry B, Hall WD: Quantification of physical illness in psychiatric research in the elderly. International Journal of Geriatric Psychiatry 1990, 5:161–170.

20. Wenger GC: A longitudinal study of changes and adaptation in the support networks of Welsh elderly over 75. Journal of Cross-Cultural Gerontology 1986, 1(3):277–304.

21. Rindfuss RR, Jampaklay A, Entwisle B, Sawangdee Y, Faust K, Prasartkul P: The Collection and Analysis of Social Network Data in Nang Rong, Thailand. In Network Epidemiology – A Handbook for Survey Design and Data Collection. Edited by: Morris M. Oxford: Oxford University Press; 2004.

22. Sherbourne CD, Stewart AL: The MOS social support survey. Soc Sci Med. 1991, 32(6):705–714.

23. Ganguli M, Chandra V, Gilby JE, Ratcliff G, Sharma SD, Pandav R, Seaberg EC, Belle S: Cognitive test performance in a community-based nondemented elderly sample in rural India: the Indo-U.S. Cross-National Dementia Epidemiology Study. International Psychogeriatrics 1996, 8(4):507–524.

24. Welsh KA, Butters N, Mohs RC, Beekly D, Edland S, Fillenbaum G, Heyman A: The Consortium to Establish a Registry for Alzheimer's Disease (CERAD). Part V. A normative study of the neuropsychological battery. Neurology 1994, 44(4):609–614.

25. Epping-Jordan J, Ustun T: The WHODAS-II: leveling the playing field for all disorders. WHO Mental Health Bulletin 2000, 6:5.

26. Stewart R, Prince M, Mann A, Richards M, Brayne C: Stroke, vascular risk factors and depression: Cross-sectional study in a UK Caribbean-born population. Br J Psychiatry 2001, 178(1):23–28.

27. Kim JM, Stewart R, Kim SW, Yang SJ, Shin IS, Yoon JS: Vascular risk factors and incident late-life depression in a Korean population. The British Journal of Psychiatry 2006, 189(1):26–30.

28. Braam AW, Prince MJ, Beekman AT, Delespaul P, Dewey ME, Geerlings SW, Kivela SL, Lawlor BA, Magnusson H, Meller I, et al.: Physical health and depressive symptoms in older Europeans. Results from EURODEP. Br J Psychiatry 2005, 187:35–42.

29. Prince MJ, Harwood RH, Blizard RA, Thomas A, Mann AH: Impairment, disability and handicap as risk factors for depression in old age. The Gospel Oak Project V. Psychol Med 1997, 27(2):311–321.

30. Prince M, Patel V, Saxena S, Maj M, Maselko J, Phillips MR, Rahman A: No health without mental health. Lancet 2007, 370(9590):859–877.

31. Boutin-Foster C: Getting to the heart of social support: A qualitative analysis of the types of instrumental support that are most helpful in motivating cardiac risk factor modification. Heart & Lung: The Journal of Acute and Critical Care 2005, 34(1):22–29.

32. Greenglass E, Fiksenbaum L, Eaton J: The relationship between coping, social support, functional disability and depression in the elderly. Anxiety, Stress and Coping 2006, 19:15–31.

33. Knodel J, Saengtienchai C, Sittitrai W: The living arrangements of elderly in Thailand: views of the populace. Journal of Cross-Cultural Gerontology 1995, 10:79–111.

34. Zimmer Z, Korniek K, Knodel J, Chayovan N: Support by migrants to their elderly parents in rural Cambodia and Thailand: A comparative study. In Poverty, Gender and Youth Working Paper no 2. New York: Population Council; 2007.

Use of Modern Contraception by the Poor is Falling Behind

Emmanuela Gakidou and Effy Vayena

ABSTRACT

Background

The widespread increase in the use of contraception, due to multiple factors including improved access to modern contraception, is one of the most dramatic social transformations of the past fifty years. This study explores whether the global progress in the use of modern contraceptives has also benefited the poorest.

Methods and Findings

Demographic and Health Surveys from 55 developing countries were analyzed using wealth indices that allow the identification of the absolute poor within each country. This article explores the macro level determinants of the differences in the use of modern contraceptives between the poor and the national averages of several countries. Despite increases in national averages, use of modern contraception by the absolute poor remains low. South and Southeast

Asia have relatively high rates of modern contraception in the absolute poor, on average 17% higher than in Latin America. Over time the gaps in use persist and are increasing. Latin America exhibits significantly larger gaps in use between the poor and the averages, while gaps in sub-Saharan Africa are on average smaller by 15.8% and in Southeast Asia by 11.6%.

Conclusions

The secular trend of increasing rates of modern contraceptive use has not resulted in a decrease of the gap in use for those living in absolute poverty. Countries with large economic inequalities also exhibit large inequalities in modern contraceptive use. In addition to macro level factors that influence contraceptive use, such as economic development and provision of reproductive health services, there are strong regional variations, with sub-Saharan Africa exhibiting the lowest national rates of use, South and Southeast Asia the highest use among the poor, and Latin America the largest inequalities in use.

Editors' Summary

Background

Access to safe and effective methods of contraception is seen by many to be a basic human right. Contraception plays an important role in improving women's health (by reducing the risks that would otherwise accompany unwanted births), as well as the social and financial situation of women and their families. However, despite a steady increase in contraceptive use worldwide over the past few decades, the World Health Organization says there is still a significant unmet need for birth control. Very many women worldwide, probably around 123 million, would like to limit the number of children that they might have but, despite this, they are not using contraception. There are probably many factors responsible for this unmet need, including the availability of health services, a woman's level of education, her social and financial situation, and cultural factors.

Why was this Study Done?

Although it is clear that use of contraception has been increasing worldwide over the past few decades, particularly in developing countries, it is not clear whether the poorest people in each country have also benefited from this trend. Given that contraception has important effects on health and on the financial and social circumstances of a family, it is important to find out whether there are any differences in contraceptive use between the poorest and richer members of society.

What did the Researchers do and Find?

This research project was based on data collected by a survey oganization about various aspects of the health, social, and economic status of households worldwide. Over 100 surveys conducted between 1985 and 2003 were used from the publicly available survey database. The researchers then classified each household for which there was survey information as being in the poorest 20% of households or not, worldwide. Importantly, this categorization reflects whether the household was in the "absolute poor" worldwide, not just the poorest for their respective country. Since information about household income was not directly available from the surveys, the researchers had to use an approach based on ownership of consumer goods and services (referred to as "asset-based wealth measures"). The researchers then looked at trends in contraceptive use amongst the poorest households, and examined whether contraceptive use was linked to other factors, such as level of education and average income.

The data showed that use of contraception by poor women was linked to the overall degree of poverty in the woman's country. Poor women from countries where many households were in the poorest 20% worldwide were far less likely to use contraception. Secondly, the researchers found that poor women were less likely to use contraception than average women in their country, and in richer countries, there seemed to be a larger gap in contraceptive use between "average" and "poor" women. Finally, the researchers found that various factors were linked to greater contraceptive use, which included the date of the survey (more recent surveys were more likely to show greater use of contraceptives), the wealth of the country where the survey was done (richer countries showed greater use), and whether women had skilled birth attendants (a marker of access to reproductive health services, and again this pointed to greater use of contraceptives). However, the researchers did find that there was huge variability in use of contraceptives worldwide, even when comparing countries at a similar economic level.

What do these Findings Mean?

This study shows that although contraception use is increasing over time, its use by poor people is low. The gap in use of contraception between poor people and "average" people also seems to be increasing over time and is wider in richer countries. The reasons behind these findings are not clear, but the data suggest that nations and international health organizations need to focus their attention on providing contraceptive services in a way that will reach people who have very low incomes.

Introduction

The use of safe and effective methods of contraception allows couples to determine the number and spacing of their pregnancies. Access to such methods was deemed a fundamental human right by the 1994 International Conference on Population and Development (ICPD)—a forum in which countries committed to work toward achieving the goal of universal access to reproductive health services, including access to effective contraceptives. Improving the use of effective contraception contributes to reducing the burden of reproductive ill health by decreasing mortality and morbidity of unwanted pregnancies [1,2]. Further, increasing contraceptive use reduces fertility, which, in turn, can play a crucial role in poverty reduction [3,4].

The widespread increase in the use of contraception is one of the most dramatic social transformations of the second half of the twentieth century [5]. Spurred by the international population control movement in the 1960s, 1970s, and 1980s, contraceptive use increased dramatically throughout the developing world [6]. This increase is likely due to multiple factors including access to modern contraception. Such access, in turn, is likely related to micro- and macroeconomic factors, including women's education, household income, integration into the modern economy, and to the proactive efforts of governments and other health providers to make contraceptive services available. A question that has not been addressed to date is whether the poor have also experienced this positive trend, which has been demonstrated for national average use rates [7,8].

The first Millennium Development Goal is the reduction in absolute poverty. The development community has increasingly focused its attention on the circumstances of the absolute poor living on $1 a day and the near-poor living on $2 a day [9]. Given the international commitment to ensuring that couples are able to exercise their right to plan their pregnancies and the important role of contraceptive use in promoting both reproductive health and economic growth, it is essential to determine whether the absolute poor in different parts of the world are able to use modern contraception. In short, is the apparent global progress in the use of modern contraceptives also benefiting the poorest?

To answer this question and provide evidence for monitoring progress toward international goals, the analysis presented here uses data from the Demographic and Health Surveys (DHS) [10], as well as a methodology identifying the absolute poor women within each country, to explore differences in progress between the rich and the poor and the macro level determinants of these differences.

Methods

Data

The findings of this analysis are based on an analysis of 110 DHS [10]. The DHS is a household survey program that collects data on maternal and child health, using nationally representative samples of women of reproductive age. Table 1 presents the countries, years, and sample sizes of the surveys used in this analysis. The surveys span a period of nearly 20 years (1985–2003) and include several countries with three or more surveys from Latin America and the Caribbean, sub-Saharan Africa, and South and Southeast Asia.

Table 1. List of Surveys Included in the Analysis

Region	Country	Year	Sample Size	Region	Country	Year	Sample Size
Latin America and the Caribbean	Bolivia	1989	7,923		Ghana	2003	5,691
	Bolivia	1994	8,603		Guinea	1999	6,753
	Bolivia	1998	11,187		Kenya	1993	7,540
	Brazil	1986	5,892		Kenya	1998	7,881
	Brazil	1996	12,612		Kenya	2003	8,195
	Colombia	1986	5,329		Liberia	1986	5,239
	Colombia	1990	8,644		Madagascar	1992	6,260
	Colombia	1995	11,140		Madagascar	1997	7,060
	Colombia	2000	11,585		Malawi	1992	4,850
	Dominican Republic	1986	7,649		Malawi	2000	13,220
	Dominican Republic	1991	7,329		Mali	1987	3,200
	Dominican Republic	1996	8,422		Mali	1996	9,704
	Dominican Republic	2002	26,000		Mali	2001	12,849
	Dominican Republic	2002	26,000		Mauritania	2000	7,728
	Ecuador	1987	4,713		Mozambique	1997	8,779
	El Salvador	1985	5,207		Namibia	1992	5,421
	Guatemala	1987	5,160		Namibia	2000	6,755
	Guatemala	1995	12,403		Niger	1992	6,503
	Guatemala	1998	6,021		Niger	1998	7,577
	Haiti	1995	5,356		Nigeria	1990	8,781
	Haiti	2000	10,159		Nigeria	2003	7,620
	Mexico	1987	9,310		Rwanda	1992	6,551
	Nicaragua	1998	13,634		Rwanda	2000	10,421
	Nicaragua	2001	13,060		Senegal	1986	4,415
	Paraguay	1990	5,827		Senegal	1993	6,310
	Peru	1986	4,999		Senegal	1997	8,593
	Peru	1992	15,882		South Africa	1998	11,735
	Peru	1996	28,951		Sudan	1990	5,860
	Peru	2000	27,840		Tanzania	1992	9,238
	Trinidad and Tobago	1987	3,806		Tanzania	1996	8,120
Southeast Asia	Bangladesh	1993	9,640		Tanzania	1999	4,080
	Bangladesh	1997	9,127		Togo	1988	3,360
	Bangladesh	1999/2000	10,544		Togo	1998	8,569
	Cambodia	2000	15,351		Uganda	1988	4,730
	India	1993	89,777		Uganda	1995	7,070
	India	1998	90,303		Uganda	2000/01	7,246
	Indonesia	1987	11,884		Zambia	1992	7,060
	Indonesia	1991	22,000		Zambia	1996	8,021
	Indonesia	1994	28,168		Zambia	2001	7,660
	Indonesia	1997	28,810		Zimbabwe	1988	4,201
	Indonesia	2002	29,483		Zimbabwe	1994	6,120
	Nepal	1996	8,429		Zimbabwe	1999	5,907
	Nepal	2001	8,726				
	Pakistan	1991	6,611				
	Philippines	1993	15,029				
	Philippines	1998	13,983				
	Sri Lanka	1987	5,865				
	Thailand	1987	6,775				
	Vietnam	1997	5,664				
	Vietnam	2002	5,665				
Sub-Saharan Africa	Benin	1996	5,491				
	Benin	2001	6,219				
	Botswana	1988	4,368				
	Burkina Faso	1992	6,354				
	Burkina Faso	1998	6,445				
	Burundi	1987	3,970				
	Cameroon	1991	3,871				
	Cameroon	1998	5,501				
	Central African Republic	1995	5,884				
	Chad	1997	7,454				
	Comoros	1996	3,050				
	Côte d'Ivoire	1994	8,099				
	Côte d'Ivoire	1998	3,040				
	Ethiopia	2000	15,367				
	Gabon	2000	6,183				
	Ghana	1988	4,488				
	Ghana	1993	4,562				
	Ghana	1998	4,843				

doi:10.1371/journal.pmed.0040001.t001

Dependent Variable

The use of contraception is analyzed for nonpregnant, ever-married women in the age group 15–49 y. We limit the analysis to ever-married women (married, divorced, or widowed), because in several countries the DHS samples only within that group. Only modern contraceptive methods are included due to their higher efficacy compared to traditional methods [11]. These methods are: injectable and oral hormones, implants, intrauterine devices, spermicides, condoms, diaphragms, female sterilization, and male sterilization. The dependent variable in the analysis is the percent of ever-married women using modern contraceptive methods, measured for each quintile of wealth in each survey.

Explanatory Variables

To examine factors that might potentially influence contraceptive use across countries and over time, regressions were run using the STATA statistical package, version 9.2 (http://www.stata.com/). Estimates of socioeconomic characteristics, skilled birth attendance, and education are directly derived from the DHS datasets, using the sampling weights. Average income per capita in international dollars is available from the Penn World Tables and the World Bank [12,13] for multiple years for each country. We used the estimate of GDP per capita for the year of the survey that was used in the analysis. The Gini index was available for 1990 and 2000; the year that was closest to the survey year was included in the study [13]. The Gini index measures the extent to which the distribution of income across households within a country is unequal; it takes on values from 0 to 100 with 0 representing perfect equality and 100 perfect inequality.

Estimation of "Wealth Quintiles"

We measured wealth using a method developed by Ferguson et al. [14]. As wealth is a latent variable that cannot be directly observed, we estimate it using information on predictors of wealth (age, education, sex of the household head, and urban/rural location) and indicators of wealth (electricity, radio, television, refrigerator; bicycle, motorcycle, car; main construction materials of the walls, roof, and floor of the house; source of drinking water and type of toilet facility, as well as, in some cases, other country-specific assets). The set of assets available in the DHS is limited, as the questionnaire was not designed to be used for the estimation of a wealth index.

A detailed discussion of the statistical methods applied to estimate the index can be found elsewhere, and its application seen in other studies [14–18]. Briefly, a random-effects probit model was used to identify "cutpoints" that represent the point on the wealth (latent) scale above which a household is more likely to own a particular asset. This "asset ladder" was then applied to every household in each survey to produce adjusted estimates of household wealth. Linear regression of asset cutpoints from all surveys was used to place the wealth estimates from all surveys on the same scale, thus leading to a wealth index that is directly comparable across surveys. The correlation between the average economic status at the country level and gross domestic product (GDP) per capita is 0.83 across all DHS surveys.

For this analysis, we have distributed the population in each country to "developing country quintiles." These quintiles have been constructed so that quintile 1 refers to the bottom 20% of the population across all developing countries, taking into account DHS sampling weights and the relative population size of each country. Thus, quintile 1 does not represent the bottom quintile in each country but can be thought of as a measure of absolute deprivation. This construction of quintiles across all surveys allows for comparisons and analyses of variations in contraceptive use rates in the most deprived.

The composition and the percentage of the population in absolute deprivation changes over time within a country, as some countries have achieved a reduction in the proportion of the population living in absolute poverty over the past two decades. Despite improvements in the levels of poverty, the gap between the national average and those in the poorest quintile in key health indicators remains of critical policy significance. A central dimension of the effectiveness of a government is its ability to deliver services to those in absolute deprivation. As the national income grows, a country's capacity to do this should increase. The construction of quintiles in a comparable way over time and across countries allowed us to monitor the use of modern contraceptive methods by the poor over time.

Regression Analysis

We explored two types of regression models: (i) ordinary least squares (OLS) regression, which was run separately for each wealth quintile and the national average; (ii) seemingly unrelated regression, which was applied to the five quintiles at the same time, to control for potential correlation of the error terms. In all regressions, the dependent variable was the percent use of modern contraceptive methods, and the independent variables were GDP per capita, Gini coefficient, year of survey, percent births attended by skilled personnel, average years of education, and a regional dummy variable.

Our findings were robust to the choice of model. In the tables and figures that follow, we chose to present the estimates from the OLS regressions, as the seemingly unrelated regressions can be applied only to countries with estimates of contraceptive use for all quintiles of income, which reduces the sample size. As the choice of model does not affect the substantive conclusions on the size and significance of the effects, we present the results from the OLS model.

Results

Average rates of use of modern contraceptive methods have increased over the past few decades in most developing countries. Our findings illustrate the differences in use by the poorest groups and how these differences relate to the estimated national average use.

Figure 1 presents a graphical exploration of the use of modern contraceptive methods by women in the poorest quintile, compared to the percentage of women in each country in that quintile. As the proportion of women in the bottom quintile increases, the estimate of use by the bottom quintile approximates the national average. Figure 1 shows, as expected, that the prevalence of modern contraceptive use by poor women is strongly related to the proportion of women in each country who are poor. The striking result from Figure 1 is that there are large variations in modern contraceptive use across countries with similar proportions of women in poverty; for example, in countries with 30% of women in the bottom quintile, the prevalence of use ranges from near zero to 24%.

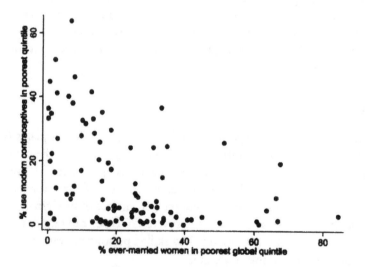

Figure 1. Modern Contraceptive-Method Use in the Poorest Global Quintile The relationship between the percentage of ever-married women and the use of modern contraceptive methods by ever-married women in the poorest global quintile is shown.

Figure 2 summarizes the relationship between the use of modern contraceptives and the level of economic development. As income per capita increases, the gap between the poorest members of the population and those representing the national average appears to increase. The positive slope in Figure 2 implies that as countries become richer, the poor remain impoverished. Among the data in Figure 2 are Mexico and Gabon (not indicated specifically), which have a comparatively high level of economic development with relatively small gaps in contraceptive prevalence between the national average and the poor. Figure 2 also highlights that at a given level of economic development, there are large variations across countries in the gaps in modern contraceptive use by the poor.

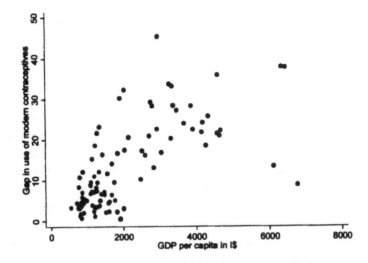

Figure 2. Modern Contraceptive Use and Level of Economic Development Gaps increase at higher incomes. The gap in modern contraceptive use rates between the national average and the poorest quintile versus GDP per capita in international dollars is shown.

To establish whether or not the poorest are being left behind in the use of modern contraceptive methods across the set of countries in our analysis, a systematic examination of the relationships is presented in Table 2, which shows the results of three multivariate regressions. The regressions use data from each survey and attempt to formalize the relationships between the use of modern contraceptives and potential macro level determinants. The three regressions explore relationships between the explanatory variables and (i) national average level of use of modern methods, (ii) use by the poorest quintile, and (iii) the gap in use between the average and the poorest.

Table 2. Results from Multivariate Regression Models

Variables included in the Model	Variable	National Average			Poorest Quintile			Difference between National Average and Poorest Quintile		
		Coefficient	Standard Error	p-Value	Coefficient	Standard Error	p-Value	Coefficient	Standard Error	p-Value
Independent variables	GDP per capita	0.003[a]	0.001[a]	0.008[a]	0.002	0.001	0.075	0.000	0.001	0.625
	Gini index	0.076	0.159	0.634	−0.128	0.144	0.379	0.206[a]	0.097[a]	0.038[a]
	Year	0.769[a]	0.220[a]	0.001[a]	0.467[a]	0.207[a]	0.027[a]	0.351[a]	0.138[a]	0.013[a]
	Skilled birth attendance[b]	0.372[a]	0.068[a]	0.000[a]	0.328[a]	0.062[a]	0.000[a]	0.062	0.043	0.151
	Education	1.125	1.184	0.345	0.816	1.146	0.479	0.642	0.720	0.376
	Constant	−1533.509	438.279	0.000	−937.062	412.488	0.026	−692.702	275.409	0.014
Regional coefficients	Latin America and the Caribbean[c]	—	—	—	—	—	—	—	—	—
	Sub-Saharan Africa	−16.990[a]	3.763[a]	0.000[a]	−4.145	3.740	0.271	−15.793[a]	2.491[a]	0.000[a]
	South and Southeast Asia	9.360[a]	4.025[a]	0.022[a]	17.864[a]	3.674[a]	0.026[a]	−11.574[a]	2.473[a]	0.000[a]
	R^2	0.78[a]	—	—	0.69[a]	—	—	0.66[a]	—	—

Multivariate regression models show (I) level of use of modern methods for national average, (II) use by the poorest quintile, and (III) the gap in use between the average and the poorest.
[a] Numbers are statistically significant at the 0.05 level.
[b] Percent of births attended by skilled personnel.
[c] Latin America and the Caribbean is the reference category, also equivalent to zero.
doi:10.1371/journal.pmed.0040031.t002

At the national level, average income per capita, year of the survey, and percent of births attended by skilled personnel are all significantly associated with higher levels of modern contraceptive use. The variable representing the year of the survey is included to capture the secular trend in use of modern contraceptives, which has been shown to be increasing over time. Skilled birth attendance can be considered as a crude proxy for access to reproductive health services, and its statistical significance implies that, controlling for all other factors in the model, it is strongly associated with higher levels of contraceptive use [19,20]. Income inequality and education of women do not seem to be significant predictors of average levels of contraceptive use. Table 2 shows that even after controlling for these macro level determinants, strong regional differences remain with South and Southeast Asian countries having the highest average levels of use, followed by Latin America and the Caribbean, and sub-Saharan Africa at the lowest average use levels.

The results for quintile 1 are markedly different from those at the national level. Skilled birth attendance and year are the only significant variables in the model, while GDP per capita and education are not. This implies that if skilled birth attendance is acting as a proxy for supply of (or access to) services, it is highly significant not only for the national average but also for the poorest populations. The coefficients on the regional effects in this regression are striking. South and Southeast Asia are the regions that show significantly higher rates of modern contraception in the absolute poor, on average 17% higher use rates than the poor in Latin America. Rates of use by the poorest quintile in Latin America and sub-Saharan Africa are not statistically distinguishable, despite significantly higher average use rates in Latin America.

The last column in Table 2 shows the results for the regression of the gap between the national average and the poorest. The year of the survey is statistically significant, suggesting that, controlling for the level of economic development, over time the bottom quintile is doing worse relative to the mean. Higher levels of income inequality are also associated with larger gaps between the national average and the poorest. The finding that countries with higher economic inequality also exhibit higher inequalities in modern contraceptive use is not surprising, but it is important, implying that inequalities in contraceptive coverage reflect overall inequalities in a country. As in the other two regressions, the regional coefficients suggest significant differences across the regions. Latin American countries exhibit significantly larger gaps in use between the poor and the average, while compared to Latin America sub-Saharan African countries have on average a gap that is smaller by 15.8% and Southeast Asian countries have on average 11.6% smaller gaps. This result implies that, even after controlling for variables that might be important in determining the use of contraception, there is a strong regional effect, with Latin America showing the largest inequalities in contraceptive use.

The R^2 coefficients shown in Table 2 suggest that the multivariate regressions explain a large amount of observed variation across countries in the national averages (78%) and high, but smaller, amounts of the differences across the poorest quintile (69%), and the gaps (66%).

Finally, Figure 3 shows the multivariate regression coefficients and 95% confidence intervals for all quintiles for skilled birth attendance and year of survey. The data in Figure 3 imply that the effect of skilled birth attendance rates is significant and at roughly the same magnitude across the bottom four quintiles. Put differently, the supply of reproductive health services provided by the health system of each country is an important determinant of rates of modern contraceptive use and is similarly important for women across levels of wealth. The only group in which the effect of skilled birth attendance is not statistically significant is women in the top wealth quintile. This suggests that national investments in reproductive health services benefit the majority of the population and are positively associated with higher rates of modern contraceptive use.

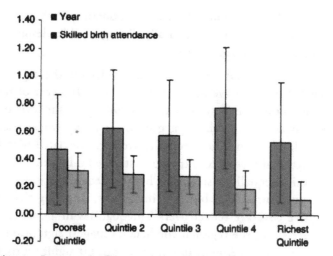

Figure 3. Multivariate Regression Coefficients and 95% Confidence Intervals for All Income Quintiles for Skilled Birth Attendance and Year of Survey

The coefficient on the year of the survey is significant for all quintiles. This variable is used as a proxy for the secular trend in contraceptive use that remains after controlling for the effect of economic development, income inequality, level of education, and urbanization. This finding suggests that the increases in contraceptive use seen over time are benefiting women in all quintiles of wealth. Furthermore, in combination with the findings presented in Table 2 that the gap between the poor and the national average is increasing over time, it suggests that the poorest women are exhibiting a slower rate of increase in the use of modern contraceptives than the rest of the population, and if this trend continues, they will continue to be left behind with regard to contraceptive use.

Discussion

Consistent with expected relationships for demand and access, the few studies undertaken to date with a focus on socioeconomic issues indicate that contraception rates are lower in poor countries and, within the limited set of countries analyzed, lower in poor women [21–25]. This study demonstrates that despite increases in national averages over time, the use of modern contraception by the absolute poor remains low, and the gaps in use across wealth quintiles persist and are increasing.

The result of this study—that the gap in modern contraceptive prevalence between the absolute poor and the rest of the population in developing countries is increasing over time and tends to widen in countries with higher incomes—needs careful exploration. Is this difference driven largely by the relationship between demand for modern contraception and economic status, or by trends and relationships related to the availability of contraceptive services, or both? At both the micro and macro levels, there is a strong relationship between modern contraception rates and economic status. This is likely due to complex pathways relating income to both the demand for and also the supply of contraceptive services. The gap in modern contraceptive use could be getting larger, because as national income per capita rises, the gap in income between the rich and the absolute poor is also rising. Given the relationship between income inequality and contraception, increasing gaps in income might be driving the increasing gaps in contraception. Another explanation of the increasing gap may be that as countries get richer, the proportion of the population living in absolute poverty is decreasing. It is possible that the composition of the poorest quintile is becoming increasingly "selective" to include the most disadvantaged and hardest-to-reach populations.

These observations, however, should not lead to complacency about accepting as inevitable that the absolute poor will always lag behind in contraceptive use. While several studies have evaluated the impact of geographical, educational, social, cultural, and political factors on the use of contraception [26–34], the findings of this analysis can be interpreted as showing that modern contraception rates in the absolute poor vary greatly across countries and are highly sensitive to the availability of services. The persistent gap between the absolute poor and the rest of the population is unlikely to be due to low demand for fertility regulation in these households. Countries could in principle differentially increase contraception rates in the absolute poor through increased provision of services that are tailored to local circumstances and financially accessible. We argue that the steadily increasing gap, in combination with greater national income inequalities, is a question of political priority for contraception and more broadly for reproductive health services.

The secular trend toward reduced levels of total fertility at any given level of national income has been well documented [35]. This reduction in fertility has been attributed in part to cultural change, as well as to changing economic and social status of women, increasing access to information, and the role of mass media. This analysis showed that for all wealth quintiles there is a statistically significant increase in contraception rates over time. This trend may be a reflection of the nexus of cultural and social transformation. While improvements are seen across all quintiles, the gap between countries' average use and use by the poor has been increasing over time. The mechanisms of cultural change, such as exposure to mass media and changing socioeconomic roles for women, may simply not be having much influence on those in absolute deprivation [30]. Regardless of the reason for the differential time trend, it implies that with each succeeding decade, inequalities in modern contraceptive use will increase unless some active policies to counteract this trend are pursued.

An important policy issue is the responsiveness of modern contraceptive prevalence to the availability or supply of contraceptive services. Careful analysis of this at the local level requires disentangling demand from a range of provider attributes including price, quality, cultural sensitivity, physical distance, and language. Direct measures of supply or availability are difficult to construct for cross-country comparisons. Skilled birth attendance rate may be considered as a crude proxy for the availability of selected reproductive heath services [36]. Attended deliveries at a given level of income will be higher where the financial, physical, cultural, and other barriers are lower. If the interpretation of skilled birth attendance as a measure of supply is valid, the results demonstrate that contraceptive prevalence is highly sensitive to the supply of services. In fact, the bottom four quintiles appear to be equally sensitive to the availability of services. This finding is consistent with analyses that have emphasized considerable unmet need for contraceptive services [31] and studies that illustrated that contraceptive use increases as more types of methods become available [37,38]. If modern contraceptive prevalence in the absolute poor can be significantly increased by enhanced supply, it highlights that the widening gap between rich and poor could be avoided through targeted interventions [39]. Improvements of the supply of services will be more effective if they take into consideration the reasons for differential uptake of modern contraception by the poor where they are available.

This study reaffirms the substantially lower levels of modern contraceptive prevalence in all income groups in sub-Saharan Africa, and to a lesser extent Latin America and the Caribbean, as compared to South and Southeast Asia [40]. Although the roles of religion, traditional concepts of family formation, health concerns, and medical barriers in demand for modern contraception have been explored [29,30,33], this analysis provides no insight into these patterns. It does, however, point out that these regional factors interact with economic status in such a way that the gaps between rich and poor are much larger in Latin America than in other regions. This gap remains even after taking into consideration supply (as approximated by skilled birth attendance), income per capita, income inequality, and secular trends. This finding highlights that Latin American health systems may need to pay particular attention to policies that affect delivery of reproductive health services to the absolute poor.

Limitations

Interpretation of these findings must take into consideration several limitations of this study. The analysis was undertaken for ever-married women only. Although in some countries modern contraceptive use by unmarried women could be substantial, this information is not available for a large number of DHS whose sampling frame includes ever-married women only, and where these data are available for unmarried women, there are concerns about the degree to which underreporting of contraceptive use due to cultural and social concerns may undermine the validity of the data. While the DHS program focuses a considerable amount of resources in making sure that the data are of high quality, the present analysis has relied on self-reported use of contraception, which suffers from several limitations, similar to other indicators measured through self-reports [41].

The wealth index has been constructed using all available indicator variables in the DHS. The DHS were not designed to measure economic status, and therefore the set of items included in this part of the questionnaire is not ideal for differentiating households throughout the distribution of income in a country. Further investments in more accurate measurement of income and identification of the absolute poor are necessary in the coming decades of reporting toward the Millennium Development Goals. A final limitation worth mentioning is that the relationships explored in this analysis are at the national level, and much could be learned by a more in-depth study of individual level determinants. Even across countries, however, a considerable amount of the variation in the use of modern contraception by the poor and the gaps with the national average use rates remains unexplained by the macro level determinants included in this analysis. Further study into factors that might explain the remaining variation could provide insight into interventions and policies that would be effective at reducing the gaps. For example, information on the level of financial commitment to family planning as a percentage of public and total expenditure on health is not available for most countries; however, it might provide insight into how the level of financial investment influences uptake of modern contraception by the poor.

Conclusions

As national income per capita rises, countries have increasing fiscal capacity to finance the delivery of services to the absolute poor. Taking advantage of this fiscal space requires, of course, the social and political commitment to use scarce resources to improve the circumstances and opportunities of the poor. This study shows that if current patterns persist, the absolute poor will continue to be left behind in the overall progress in increasing modern contraceptive use. On the positive side, it appears that making reproductive health services available is a powerful determinant, all other things being equal, of contraceptive prevalence. A concerted effort by governments to facilitate an increase in physical, financial, and cultural access to reproductive health services for the poor could have a major effect. The fundamental challenge will be to raise the international and national priority accorded to reproductive health services for the poor. Paradoxically, in an era of increased resource flows for global health through mechanisms like the Global Fund for AIDS, Tuberculosis and Malaria and GAVI (the Global Alliance for Vaccines and Immunisations), contraceptive use and related reproductive health services seem increasingly difficult to place on the health agenda [42–44]. The trends that have been observed to date provide strong evidence that without new priority attention to modern contraception, the poor will remain deprived of the fundamental right to its demonstrated benefits.

Acknowledgements

We thank Herbert Peterson for providing comments at various stages of the manuscript and Metin Gulmezoglu for comments on the final draft. The authors wish to thank Johanna Riesel and Diana Lee for research assistance.

Authors' Contributions

EG and EV developed the idea, conducted the literature reviews, conducted the data analysis, and contributed to the manuscript. EV is a staff member of the World Health Organization. She alone is responsible for the views expressed in this publication, and they do not necessarily represent the decisions, policy, or views of the World Health Organization.

References

1. Collumbien M, Gerressu M, Cleland J (2004) Non-use and use of effective methods of contraception. In: Ezzati M, Lopez AD, Rodgers A, Murray CJL, editors. Comparative quantification of health risks: Global and regional burden of disease attributable to selected major risk factors. Geneva: World Health Organization. pp. 1255–1319.

2. Marston C, Cleland J (2004) The effects of contraception on obstetric care. Geneva: World Health Organization. Available: http://www.who.int/reproductive-health/publications/2004/effects_contraception/index.html. Accessed 29 November 2006.

3. Bloom D, Canning D (2005) Population, poverty reduction, and the Cairo Agenda. Proceedings of the Seminar on the Relevance of Population Aspects on the Achievement of the Millennium Development Goals. New York: United

Nations Population Fund (UNFPA). Available: http://www.un.org/esa/population/publications/PopAspectsMDG/PopAspects.htm. Accessed 29 November 2006.

4. United Nations Population Fund (2005) Reducing poverty and achieving the Millennium Development Goals: Arguments for investing in reproductive health and rights. New York: United Nations Population Fund (UNFPA). Available: http://www.unfpa.org/publications/detail.cfm?ID=243. Accessed 29 November 2006.

5. Rosenfield A, Schwartz K (2005) Population and development—Shifting paradigms, setting goals. N Engl J Med 352: 647–649.

6. Weinberger MB (1994) Recent trends in contraceptive use. Popul Bull UN 36: 55–80.

7. Gwatkin DR (2000) Health inequalities and the health of the poor: What do we know? What can we do? Bull World Health Organ 78: 3–18.

8. Victora CG, Wagstaff A, Schellenberg JA, Gwatkin D, Claeson M, et al. (2003) Applying an equity lens to child health and mortality: More of the same is not enough. Lancet 362: 233–241.

9. United Nations Development Programme (2006) Millennium Development Goals. New York: United Nations Development Programme. Available: http://www.undp.org/mdg/goalxgoal.shtml. Accessed 29 November 2006.

10. Measure DHS (2006) Demographic and health surveys. Calverton (Maryland): Measure DHS. Available: www.measuredhs.com. Accessed 29 November 2006.

11. Steiner M, Trussell J, Mehta N, Condon S, Subramaniam S, et al. (2006) Communicating contraceptive effectiveness: A randomized controlled trial to inform a World Health Organization family planning handbook. Am J Obstet Gynecol 195: 85–91.

12. Heston A, Summers R, Aten B (2002 October) Penn World Table Version 6.1. Philadelphia: Center for International Comparisons at the University of Pennsylvania (CICUP). Available: http://pwt.econ.upenn.edu/php_site/pwt_index.php. Accessed 29 November 2006.

13. World Bank (2005) World development indicators 2005. Washington (D. C.): International Bank for Reconstruction and Development/The World Bank. Available: http://devdata.worldbank.org/wdi2005/index2.htm. Accessed 29 November 2006.

14. Ferguson BD, Tandon A, Gakidou E, Murray CJL (2003) Estimating permanent income using indicator variables. In: Murray CJL, Evans DB, editors.

Health systems performance assessment: Debates, methods, and empiricism. Geneva: World Health Organization. pp. 747–760. Available: http://www.who.int/publications/2003/hspa/en/. Accessed 29 November 2006.

15. Gakidou E, Lozano R, Gonzalez-Pier E, Abbott-Klafter J, Barofsky J, et al. (2006) Assessing the effect of the 2001–2006 Mexican health reform: An interim report card. Lancet 368: 1920–1935.

16. Lozano R, Soliz P, Gakidou E, Abbott-Klafter J, Feehan D, et al. (2006) Benchmarking of performance of Mexican states with effective coverage. Lancet 368: 1729–1741.

17. Pongou R, Salomon J, Ezzati M (2006) Health impacts of macroeconomic crises and policies: Determinants of variation in childhood malnutrition trends in Cameroon. Int J Epidemiol 35: 648–656.

18. Pongou R, Ezzati M, Salomon JA (2006) Household and community socioeconomic and environmental determinants of child nutritional status in Cameroon. BMC Public Health 6: 98.

19. Adam T, Lim SS, Mehta S, Bhutta Z, Fogstad H, et al. (2005) Cost effectiveness analysis of strategies for maternal and neonatal health in developing countries. BMJ 331: 1107.

20. Mexican Ministry of Health (2006) Effective coverage of the health system in Mexico 2000–2003. Mexico, DF: Ministry of Health. Available: http://www.globalhealth.harvard.edu/pop_health_metrics5.html. Accessed 29 November 2006.

21. Clements S, Madise N (2004) Who is being served least by family planning providers? A study of modern contraceptive use in Ghana, Tanzania and Zimbabwe. Afr J Reprod Health 8: 124–136.

22. Montagu D, Prata N, Campbell MM, Walsh J, Orero S (2005) Kenya: Reaching the poor through the private sector—A network model for expanding access to reproductive health services. Health Nutrition and Population Discussion Paper. Washington (D. C.): International Bank for Reconstruction and Development/The World Bank. Available: http://web.worldbank.org/. Accessed 29 November 2006.

23. Onwuzurike BK, Uzochukwu BS (2001) Knowledge, attitude and practice of family planning amongst women in a high density low income urban of Enugu, Nigeria. Afr J Reprod Health 5: 83–89.

24. Schoemaker J (2005) Contraceptive use among the poor in Indonesia. Int Fam Plan Perspect 31: 106–114.

25. World Bank (2004) Conditions among the poor and the better-off in 56 countries. Round II country reports on health, nutrition, and population.

Washington (D. C.): World Bank. Available: http://web.worldbank.org/ Accessed 29 November 2006.

26. Ainsworth M, Beegle K, Nyamete A (1996) The impact of female schooling on fertility and contraceptive use: A study of fourteen sub-Saharan countries. Washington (D. C.): World Bank Development Report. pp. 85–122.

27. Ali M, Cleland J (1995) Contraceptive discontinuation in six developing countries: A cause-specific analysis. Int Fam Plan Perspect 21: 92–97.

28. Bongaarts J, Bruce J (1995) The causes of unmet need for contraception and the social content of services. Stud Fam Plann 26: 57–75.

29. Bongaarts J (2003) Completing the fertility transition in the developing world: The role of educational differences and fertility preferences. Popul Stud (Camb) 57: 321–335.

30. Bongaarts J (2006) The causes of stalling fertility transitions. Stud Fam Plann 37: 1–16.

31. Casterline J, Sinding S (2000) Unmet need for family planning in developing countries and implications for population policy. Popul Dev Rev 26: 691–723.

32. Cebeci SD, Erbaydar T, Kalaca S, Harmanci H, Cali S, et al. (2004) Resistance against contraception or medical contraceptive methods: A qualitative study on women and men in Istanbul. Eur J Contracept Reprod Health Care 9: 94–101.

33. Shah I (2001) Perspectives of users and potential users on methods of fertility regulation. In: Puri CP, Van Look PFA, Sachdeva G, Penhale C, editors. Sexual and reproductive health recent advances, future decisions. New Delhi: New Age International (P) Limited, Publishers. pp. 45–92.

34. Welsh MJ, Stanback J, Shelton J (2006) Access to modern contraception. Best Pract Res Clin Obstet Gynaecol 20: 323–338.

35. United Nations Population Division (2005) World contraceptive use 2005. New York: United Nations. Available: http://www.un.org/esa/population/publications/contraceptive2005/WCU2005.htm. Accessed 29 November 2006.

36. Ahmed S, Mosley WH (2002) Simultaneity in the use of maternal-child health care and contraceptives: Evidence from developing countries. Demography 39: 75–93.

37. Herndon N (1993) Next step for Egypt—Access to more methods. Network 13: 18–21.

38. Shelton JD, Davis SS (1996) Some priorities in maximizing access to and quality of contraceptive services. Adv Contracept 12: 233–237.

39. Freedman LP, Waldman RJ, de Pinho H, Wirth ME, Chowdhury AM, et al. (2005) Transforming health systems to improve the lives of women and children. Lancet 365: 997–1000.

40. Rutstein SO, Kiersten J (2004) The DHS wealth index. DHS Comparative Reports number 6. Calverton, MD: ORC Macro. Available: http://www.measuredhs.com/pubs/pub_details.cfm?ID=470. Accessed 29 November 2006.

41. World Health Organization (2006) Reproductive health indicators—Guidelines for their generation, interpretation and analysis for global monitoring. Geneva: World Health Organization. 67 p.

42. Burke AE, Shields WC (2005) Millennium Development Goals: Slow movement threatens women's health in developing countries. Contraception 72: 247–249.

43. Hwang AC, Stewart FH (2004) Family planning in the balance. Am J Public Health 94: 15–18.

44. Stewart FH, Shields WC, Hwang AC (2004) Cairo goals for reproductive health: Where do we stand at 10 years? Contraception 70: 1–2.

Copyrights

Index